THE KIDS' FIELD GUIDE

to Birds

Para mi mami.
Thank you for buying me birdseed at Christmas.

And for my dad.
Though you didn't get to see this, I know you would have been excited to read it.

Quarto.com

© 2024 Quarto Publishing Group USA Inc.
Text © 2024 Daisy Yuhas

First Published in 2024 by Cool Springs Press,
an imprint of The Quarto Group,
100 Cummings Center, Suite 265-D, Beverly, MA 01915, USA.
T (978) 282-9590 F (978) 283-2742

Cool Springs Press titles are also available at discount for retail, wholesale, promotional, and bulk purchase. For details, contact the Special Sales Manager by email at specialsales@quarto.com or by mail at The Quarto Group, Attn: Special Sales Manager, 100 Cummings Center, Suite 265-D, Beverly, MA 01915, USA.

28 27 26 25 24   1 2 3 4 5

ISBN: 978-0-7603-8561-6

Digital edition published in 2024
eISBN: 978-0-7603-8562-3

Library of Congress Cataloging-in-Publication Data

Names: Yuhas, Daisy, author.
Title: The kids' field guide to birds : 80+ species profiles : how to get
   started : activities and fun facts / Daisy Yuhas.
Description: Beverly, MA : Cool Springs Press, 2024. | Includes index. |
   Audience: Ages 9-11 Years | Summary: "The Kids' Field Guide to Birds is
   an exciting introduction to bird spotting for kids (and their parents),
   helping to identify common species while keeping things fun with
   activities and features throughout"—Provided by publisher.
Identifiers: LCCN 2023047230 | ISBN 9780760385616 (trade paperback) | ISBN
   9780760385623 (ebook)
Subjects: LCSH: Bird watching—Juvenile literature. |
   Birds—Identification—Juvenile literature.
Classification: LCC QL677.5 .Y84 2024 | DDC 598.072/34—dc23/eng/20231107
LC record available at https://lccn.loc.gov/2023047230

Design and Page Layout: Megan Jones Design
Cover Image: Images: From top to bottom, Traci Sepkovic, Judd Patterson, Deborah Bifulco, Traci Sepkovic, Deborah Bifulco, MiaDora McPherson
Illustration: Mattie Wells Design, page 15

Printed in China

THE KIDS' FIELD GUIDE

to Birds

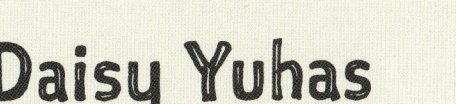

80+ SPECIES PROFILES

~~~~~~

HOW TO GET STARTED

~~~~~~

ACTIVITIES AND FUN FACTS

Daisy Yuhas

COOL
SPRINGS
PRESS

CONTENTS

Urban Birds: City Sights 16

Garden, Park, and Feeder Visitors: Birds and Blooms 27

Lake, River, and Marsh Birds: Freshwater Fans 48

INTRODUCTION

I started birding through my bedroom window when I was 14. I remember the first time I looked up a bird in a guidebook. It was a snowy winter morning. I woke up and I noticed two little gray birds sitting in the bare branches of a pear tree. One was charcoal—dark with a white beak and belly. The other was a lighter silver with a peaked cap of feathers. Each was puffed up against the cold.

I just so happened to have the *National Audubon Society Nature Guides Eastern Forests* nearby. I'd needed the book for a school project. Curious, I thumbed through photos, looking for my feathered visitors. I'd never done this before, so the search took me some time. I leafed past Olive-sided Flycatchers and Blackpoll Warblers. Finally, I spotted my birds: a Dark-eyed Junco and Tufted Titmouse. I was thrilled.

My hope is that this book brings that same sense of delight and discovery for many other readers. This field guide is a little bit different from classic, comprehensive, or regional guides. Rather than capture every species in detail, these pages document a selection of the most common and iconic birds of North America—with particular emphasis on birds you're likely to see in many parts of the continental United States and Canada. The result is an entryway for exploring a whole world of birds.

HOW TO USE THIS BOOK

The guide is organized by the places where you are likely to encounter certain birds. Be aware that some species are very adaptable, meaning you may find them in more than one kind of environment.

For each species, there is a brief description to help you identify a given bird:

- **Appearance** points to key identifying features.
- **Approximate Length** estimates an adult bird's measurements from the tip of their beak or bill to the end of their tail.
- **Voice** flags a notable call or song— or describes the bird's repertoire.
- **Conservation Status** includes some information on whether this species' population is declining or otherwise of concern to environmental experts.
- **Location** refers to habitats and regions where this bird is commonly observed in North America. (Some species may be found in other parts of the world too—and several birds in this book travel to other regions, such as South America, each year.)
- **Diet** lists some of this bird's favorite meals— but is not a complete menu.

Throughout the book you'll also find activities and ideas to help you learn more about birds, birding, and the natural world!

CAN'T IDENTIFY A BIRD?

If you are trying to match a species you see to one in this book, there are several steps you can take. First, make some notes about the bird you want to identify. You can even snap a photo or draw a picture. Then, try to find the closest match to that bird (there may be more than one).

Read the full descriptions carefully. For some species, a photo can be deceptive. Some birds look very different based on whether they are male, female, adult, or young. Some even vary in color in different parts of the country!

Feeling stumped? If you are not finding a clear match, it could be time to look at a larger or more specialized guidebook. On page 93, you'll find a list of resources including apps, books, and websites that can help.

BIRDS AND CLIMATE CHANGE

Range maps are a traditional way of sharing where species live and migrate. The best range maps are online (see Resources, page 93) and, unlike print maps, are updated frequently. That's important because bird ranges are changing rapidly.

Human activities—like cutting forests, creating electricity, and driving cars—are releasing greenhouse gases into the air that change our planet's climate. Over time, the buildup of these greenhouse gases in the atmosphere is warming our planet. That extra heat transforms ecosystems around the world. As regions experience dramatic change, some species move to find new territory they are better suited for. In one study, scientists looked at 90 years of bird counts and discovered that species from every habitat group were changing their winter territory.

The consequences of these changes are serious. Researchers from the National Audubon Society studied records from 604 North American bird species and found that 389 species may be at risk of extinction because of our rapidly changing climate. Fortunately, those same scientists found that taking action to slow climate change could give many of these birds the time they need to adapt.

HEY, GROWN UPS!

This book is designed to help you learn alongside kids. Many of the activities in this book benefit from an adult helper. Together you will learn how to create a garden to attract hummingbirds, select a birdhouse for Wrens, or join a community science project.

GETTING INVOLVED

You can also support the budding birder in your life by getting involved in local and national organizations that protect wildlife and their habitat. Many of these groups organize programs and learning opportunities. They can also be a fantastic resource for finding ways to discover and support the species nearest you.

WELCOMING PEOPLE AND WILDLIFE

If you have access to spaces for watching or listening to birds, there are several things you can do to make these areas special. From an environmental perspective, you can support efforts that protect shared parks and sanctuaries. If you keep a personal or community garden, choose native plants and think about landscaping that reflects the local climate and ecology.

We should also consider human visitors. People of all backgrounds can take joy in birds. Consider who in your community has easy access to nature—and whether you can invite others to share those areas. Organizing a group trip to a park with other kids and parents, for instance, could be a start. Think in advance about how to include people with disabilities on bird outings. Some trails, for example, are more accessible to people who use walkers or wheelchairs. You can also de-emphasize birding by sight and practice listening to bird calls together.

Talk to your kids about the fact that birding is for everyone. Before you go outside, brainstorm together how to be friendly and respectful towards others. All are welcome.

BIOLOGY, CONSERVATION, AND BEYOND

Climate change, pollution, pesticides, and other threats put wildlife in danger. Since 1970, North America has lost nearly 30 percent of its bird species. Recent research shows many common birds are in decline. That means even species we see often in certain parts of the continent could someday disappear.

But everyone can take steps to help. Throughout this book you will discover conservation stories that reflect on the difference people can make when they care about their environment and their animal neighbors.

These pages offer an introduction to concepts from ecology and animal behavior. With a better understanding of birds, we can help them thrive.

A NOTE ABOUT NAMES

As this book moved to publication, the American Ornithological Society began an important new project: renaming several species. Rather than name birds after people, the group wants species names to reflect aspects of the animal being named, such as a bird's appearance or behavior. Five species mentioned in this book, for example, could have new names in the years to come!

BIRDING BASICS

To get started, the most essential thing you need is curiosity. Birding is all about a willingness to observe the world around you, whether it's through a window or on a dedicated walk through your neighborhood.

To make the most of your birding experience, the second thing you'll want is a field guide—like this book! You can also use an app or website (see page 93). When you notice a bird, you can then consult your guides to match the species that you see or hear.

Last but not least, many birders find it useful to keep some extra tools—like binoculars, a pencil and paper, or even a camera—to document what they observe. You can take notes to record where and when you spotted a bird, for example. That information can help you find that species again. You can also draw pictures or take a photo if you see a bird but don't have your guidebook nearby. With an image or description, you can find the species later.

I SEE A BIRD! WHAT NOW?

Every time you spot a species for the first time, you get to play detective. So as you observe birds, use your senses to take in as many clues as possible.

Here are a few questions to get you started:

◆ Where are you—and what kind of habitat are you observing?

◆ What time of year is it?

◆ How big do you think that bird is? Smaller than your hand? Bigger than a beachball?

◆ Can you describe the bird's color and shape—or even its song?

Using those details, consult your guidebooks to find a specific species. As you practice, you may notice more and more details. For instance, you can start to pay attention to bird behaviors, such as when a Northern Mockingbird flashes its wings, revealing large white patches, or when Cedar Waxwings pass berries back and forth. You might also learn more about habitats and sensory clues, such as the distinctive smells in places with seabirds nearby or the varied nesting options available in wooded areas.

HOW TO USE BINOCULARS

If there's one tool that can upgrade your bird-ing experience, it's a good set of binoculars. A pair listed as having 8 × 40 magnification can be especially handy and binoculars that come with a strap are best. Keep it looped around your neck and arm for easy access.

Before you try them with a bird, practice using a familiar object or landmark that won't fly away. Here's how:

1. First pick your target, such as a lamppost or sign. Look at it without binoculars.

2. Hold your head and gaze in that position while you lift up the binoculars to your eyes.

3. You should see the same view—though it may be blurry. If you're seeing everything in miniature or with a lot of blackness, try reversing which side of the binoculars you look through. You may also be able to twist the eyecups on your binoculars so they sit more comfortably against your face.

4. Gently fold or open the binoculars with your hands on either side to adjust the distance between the binocular lenses. (The idea is to get the lenses lined up to fit the spacing of your eyes.)

5. Slowly turn the center dial to make the image sharper or fuzzier. Experiment a little bit! When you have a clear view of your target, then the binoculars are in focus and ready to go.

PRO TIP

If you are having trouble focusing, take a careful look at your binoculars. Many have both a focus wheel in the middle *and* a smaller wheel on one side—close to where your binoculars touch your eyes—that you can fine-tune to suit your vision.

WANT TO SEE MORE BIRDS?

If you're birding from your window, you might be able to set up a feeder, a bird-friendly container plant, or even a birdbath to attract visitors. You'll find ideas for each of those projects later in this book (see pages 40, 44, and 46). Just remember that these items need to be cleaned and well-kept to be safe for feathered friends.

If you're heading outdoors, you can take several steps to spot more species. For example, you might visit a location with more than one kind of habitat—like a park with both trees and water features. Another tip: When you're outside and looking for birds, remember that loud noises, bright clothing, and pets may scare wildlife away. Visit your favorite park at more than one time of day to see if there are especially quiet, calm times.

BE PATIENT!

Sometimes, the secret to spotting birds is a few extra seconds of stillness and silence. In fact, some people use birding as a form of meditation because it is a way to pause and relax. Don't be discouraged if you don't see anything right away—you might even set a 5-minute timer to practice keeping still at your favorite birding location.

MAP OF NORTH AMERICA

More than 2,000 bird species make the North American continent home. While some of these birds are very flexible in their choice of habitat others are highly specialized. This map offers an orientation of some of the key environments found across this landmass.

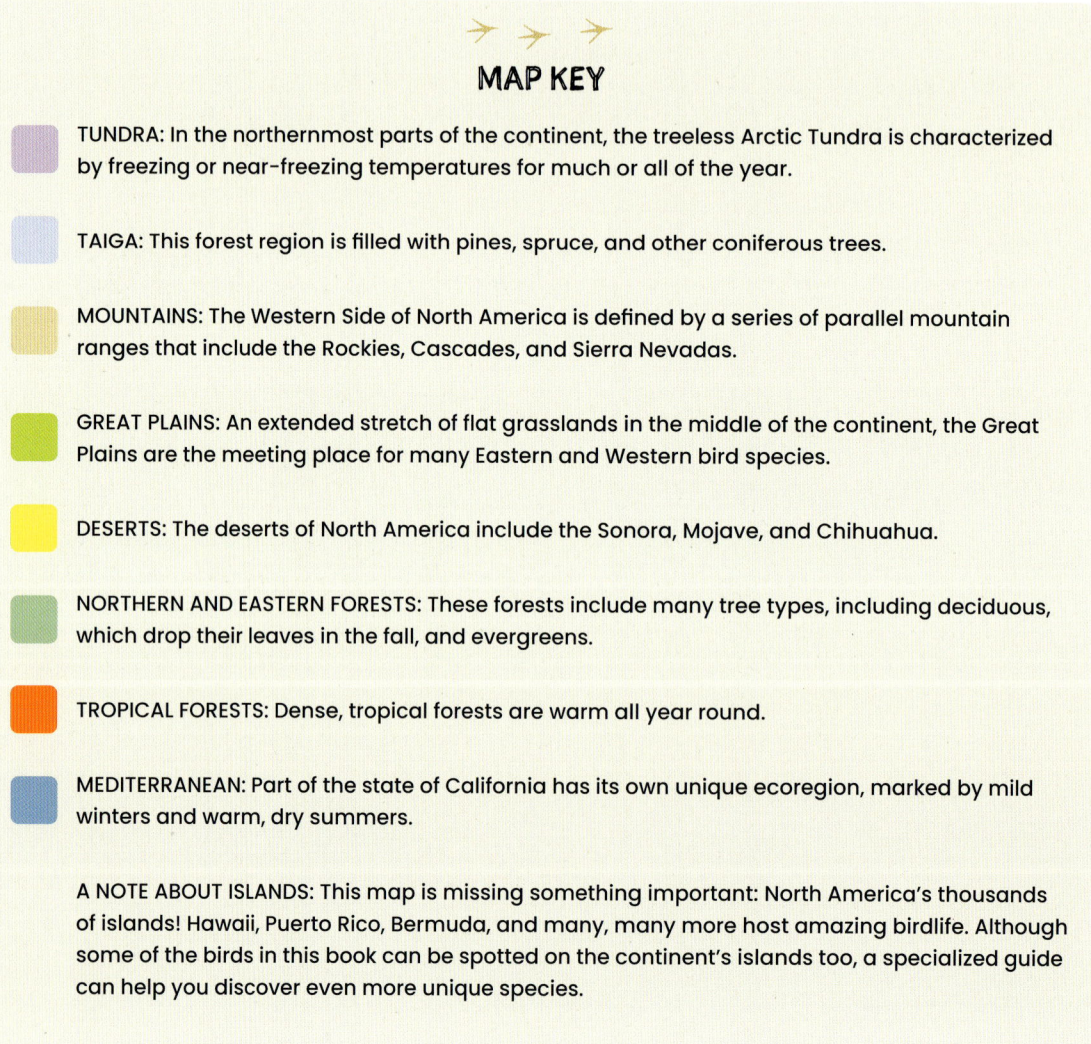

MAP KEY

TUNDRA: In the northernmost parts of the continent, the treeless Arctic Tundra is characterized by freezing or near-freezing temperatures for much or all of the year.

TAIGA: This forest region is filled with pines, spruce, and other coniferous trees.

MOUNTAINS: The Western Side of North America is defined by a series of parallel mountain ranges that include the Rockies, Cascades, and Sierra Nevadas.

GREAT PLAINS: An extended stretch of flat grasslands in the middle of the continent, the Great Plains are the meeting place for many Eastern and Western bird species.

DESERTS: The deserts of North America include the Sonora, Mojave, and Chihuahua.

NORTHERN AND EASTERN FORESTS: These forests include many tree types, including deciduous, which drop their leaves in the fall, and evergreens.

TROPICAL FORESTS: Dense, tropical forests are warm all year round.

MEDITERRANEAN: Part of the state of California has its own unique ecoregion, marked by mild winters and warm, dry summers.

A NOTE ABOUT ISLANDS: This map is missing something important: North America's thousands of islands! Hawaii, Puerto Rico, Bermuda, and many, many more host amazing birdlife. Although some of the birds in this book can be spotted on the continent's islands too, a specialized guide can help you discover even more unique species.

This map offers a simplified look at some of North America's ecological regions.

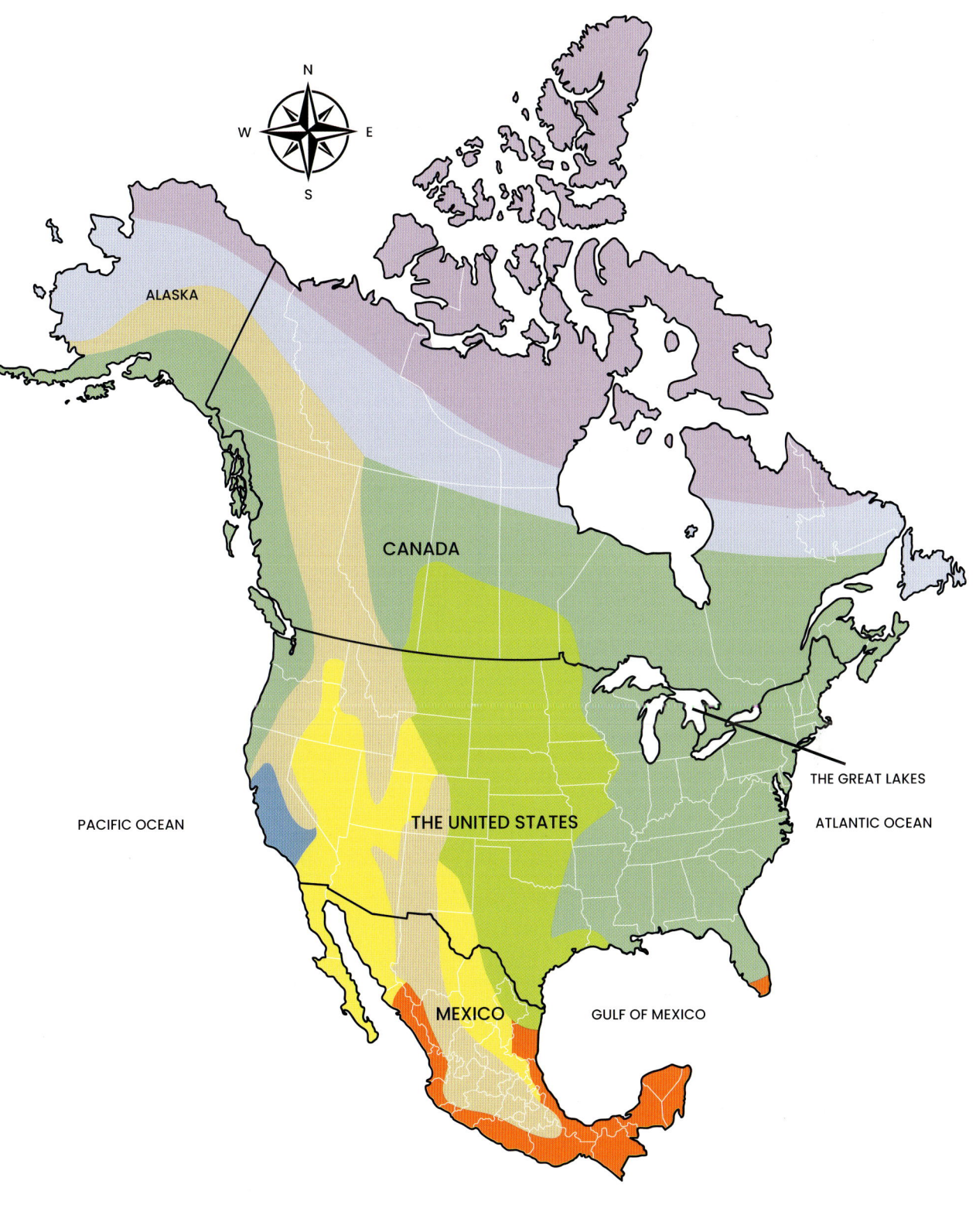

ALASKA

CANADA

PACIFIC OCEAN

THE UNITED STATES

THE GREAT LAKES

ATLANTIC OCEAN

MEXICO

GULF OF MEXICO

Urban Birds:
City Sights

AMERICAN CROW

Corvus brachyrhynchos

APPEARANCE: These all-black birds have a glossy sheen as adults. Compared with Ravens (page 84), they have a slimmer beak, smaller body, and squared-off tail feathers.

APPROXIMATE LENGTH: 17" (43 cm)

VOICE: "Caw! Caw! Caw!"

CONSERVATION STATUS: These abundant birds are not a species of concern.

LOCATION: Across much of the United States and Canada, except for deserts and Arctic areas

DIET: A little bit of everything: seeds, nuts, shellfish, reptiles, fruit, human leftovers, and even chicks from other birds

AMERICAN CROW

ABOUT

The amazing, adaptable American Crow can be found on fields, farms, waterways, woodlands, towns, and cities throughout much of the United States and Canada. Some people have treated these birds as pests, trying to bring down their numbers through hunting and other tactics. But clever Crows continue to thrive. Thousands may gather together to rest in the same area—called a roost—overnight.

One secret to success is their flexibility. Not only do they eat a wide variety of foods, they are also good problem solvers. They can figure out how to open trash bags and lunchboxes, for instance. They also have amazing memories. Scientists have found that Crows can recognize human faces and will remember how people treated them in the past. These brainy birds will play with friendly pet dogs, memorize garbage truck routes, and collect shiny trinkets. Much like our own species, Crows use their varied skills to flourish in many environments.

The American Crow is a very social species. They probably owe some of their smarts to the ways they learn from one another. These birds often stick close to their family, with many generations living together. And when a Crow dies, many will gather to inspect the body, perhaps to investigate the cause of death.

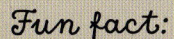

Fun fact:
Some Crows remain around their parents' nest for several years, helping to raise younger siblings.

AMERICAN ROBIN

Turdus migratorius

APPEARANCE: This familiar bird has a gray-brown back and a bright red-orange breast. Their heads are often dark with a streaked white throat patch. Their bills are bright yellow.

APPROXIMATE LENGTH: 10" (25 cm)

VOICE: The Robin's tuneful singing sounds a little bit like it's saying "Cheerily, cheer up, cheerio!"

CONSERVATION STATUS: An abundant, widespread bird

LOCATION: The American Robin is found throughout North America. Some of these birds breed in Alaska and Canada in summer, then winter over in Mexico and the southern United States.

DIET: Mostly earthworms and insects, occasionally other treats such as berries and spiders

ABOUT

Across the continent, American Robins are highly skilled at living alongside humans. They forage for food on grassy lawns and collect berries along gardens and hedges. In fact, they are an important predator for earthworm species brought to North America from Europe hundreds of years ago, which can damage ecosystems. (Learn more about native and nonnative species on page 20.)

People can help American Robins thrive by avoiding pesticides, which can harm these birds.

AMERICAN ROBIN

BIRDING TIP

Figuring out a bird's size can be tricky. You're probably not going to be close enough to get out a ruler and check their approximate length! But American Crows and American Robins are two great examples of reference birds. If you see another bird, ask yourself if it's smaller or bigger than these species to better estimate its size.

Fun fact:
The American Robin's light blue eggs may have that color as a kind of sun protection, letting eggs get warm without overheating.

CEDAR WAXWING

Bombycilla cedrorum

APPEARANCE: These sleek birds have soft cinnamon-coloration on their face and crest. Their black face mask is trimmed in white and their bellies are buttery yellow. Their wings and tail feathers are largely gray-brown and their tail tips are bright yellow or orange. The name "Waxwing" refers to small waxy red tips on their wings.

APPROXIMATE LENGTH: 6″ (15 cm)

VOICE: A faint, high-pitched whistle

CONSERVATION STATUS: Overall, this bird's numbers may be increasing.

LOCATION: Across the continent, except Arctic areas

DIET: Fruits and berries are favorites, along with sweet sap, flowers, and insects in summer

ABOUT

Cedar Waxwings adore eating fruits. Whole flocks will gather on fruiting and flowering trees and bushes. They can gulp berries whole. They even have social rituals around sharing sweet treats. As part of courtship, which is how birds select mates, Cedar Waxwings will pass a berry back and forth, a bit like offering each other the first bite of a meal. Their diet can change this bird's colors. When these birds eat berries from a honeysuckle plant, their yellow-tipped tails can turn orange.

CEDAR WAXWING

CHIMNEY SWIFT

Chaetura pelagica

APPEARANCE: Mostly gray-brown; these birds have a distinct silhouette in flight, with narrow, long wings and a slim, torpedo-like body.

APPROXIMATE LENGTH: 5″ (13 cm)

VOICE: High-pitched chitters

CONSERVATION STATUS: Vulnerable to extinction with numbers in serious decline

LOCATION: These birds breed across the eastern United States and winter in South America

DIET: Beetles, flies, and other flying insects

ABOUT

The Chimney Swift's body is perfect for long distance, high-speed flying. They can catch thousands of insects daily and do almost everything from the air. They can even take a bath mid-flight! Their feet and tail help them cling to walls, allowing them to nest inside stone or brick chimneys or large, hollow trees. To support these birds, some people build special nesting towers just for Swifts.

THE UNDERSIDE OF A CHIMNEY SWIFT IN FLIGHT

EUROPEAN STARLING

Sturnus vulgaris

APPEARANCE: In winter, these iridescent black birds are speckled with white-tipped feathers that look like stars against a night sky, a pattern that may have earned this bird its name. In summer, they are more solidly dark, which causes their orange-red legs and yellow beaks to stand out.

APPROXIMATE LENGTH: 8.5" (22 cm)

VOICE: This bird can whistle, squeak, and mimic the calls of other animals as well as human machinery.

CONSERVATION STATUS: Widespread and common—millions of European Starlings live in North America, and they are sometimes seen as a pest because they compete with other species and consume human crops.

LOCATION: Most of North America except the High Arctic

DIET: Insects, such as caterpillars and beetles, but also fruits and seeds

EUROPEAN STARLING

ABOUT

Sometime in the late 1800s, people brought these star-spangled birds to North America from Europe. Not everyone agrees on exactly how it happened—some stories suggest Shakespeare enthusiasts let European Starlings loose in Central Park—but once these birds got here, they thrived. Starlings are impressive in many respects. They can copy other birds' calls and they sometimes gather in massive flocks to fly together in clouds called "murmurations."

WHAT ARE "NATIVE" AND "NONNATIVE" BIRDS?

Scientists call plants and animals "native" when they live in the environment where they have evolved—fitting a very particular role in that habitat. When people transplant a species to new areas—whether it's new continents or even new regions—the relocated species doesn't have the same relationship with their surroundings as a species that evolved in that area.

The results can be complicated. Some transplanted species struggle in a new environment. Others, such as the Monk Parakeet (page 24), seem to thrive without causing much disruption. But with still other species, such as the European Starling, some scientists worry they may displace local birds by competing for food and nesting sites. Ironically, European Starlings are booming in North America but declining in their native Europe, due to habitat loss and transformation there.

GREAT-TAILED GRACKLE

MALE GREAT-TAILED GRACKLE

Quiscalus mexicanus

APPEARANCE: Females and young birds are dark brown and slender. Adult males are slender, with long V-shaped tails, striking yellow eyes, and iridescent black feathers.

APPROXIMATE LENGTH: 16″ (41 cm)

VOICE: Chattering, clucking, and whistling, sometimes sounding mechanical

CONSERVATION STATUS: Low concern—in fact, their range is increasing

LOCATION: Abundant in the southwestern United States and Mexico

DIET: A little bit of everything, including spiders, fish, lizards, other birds—they love human foods too

ABOUT

Great-tailed Grackles swagger, strut, and sing like the coolest guests at a party. They gather in massive groups, making a ruckus from rooftops and streetlamps before settling for the night. Their complex calls have fascinated humans for centuries. According to traditional stories from Mexico, the *zanates* or grackles could not speak until they stole seven songs from the sea turtle, each representing a different passion, such as fear, courage, or love. These birds divide opinion. Though beloved by Aztec emperors in Central America, some folks today call a group of grackles "an annoyance."

HOUSE FINCH

MALE HOUSE FINCH

Haemorhous mexicanus

APPEARANCE: Young males and adult females are streaky brown and white all over. Adult males have rosy red feathers on their breast and head.

APPROXIMATE LENGTH: 5″ (13 cm)

VOICE: Chirps and warbles

CONSERVATION STATUS: Abundant, widespread birds

LOCATION: Common across most of the United States and Mexico

DIET: Buds, berries, and seeds are the bulk of their diet, along with occasional insects

ABOUT

A male House Finch, with its cheery cherry feathers, is a memorable sight. Once native to the western half of the continent, they spread east in the early twentieth century, when pet owners released these little birds.

This bird is often confused with the **Purple Finch**, found along the Pacific Coast and eastern half of the United States—as well as much of Canada in the summer. Male Purple Finches are pinker while House Finches are more orange. Female Purple Finches have a white eyebrow-like stripe that House Finches lack.

HOUSE SPARROW

Passer domesticus

APPEARANCE: Adult males have gray undersides and brown, black, and white streaks on top. They have a black patch on their chin and chest. In rural areas, males have white cheeks, a rust-colored neck and gray crown. Females are a drab gray-brown underneath with tan, brown, and black striping above.

APPROXIMATE LENGTH: 6" (15 cm)

VOICE: "Cheep-cheep-cheep-cheep"

CONSERVATION STATUS: Common and widespread

LOCATION: Abundant around human towns and cities throughout much of Canada, the United States, and Mexico

DIET: Seeds, grains, insects, and breadcrumbs left behind by humans

MALE HOUSE SPARROW

ABOUT

Sparrows have a long history with humanity. Archaeologists have found evidence that birds much like the House Sparrow have been living alongside humans—and feasting on our crumbs—for at least 100,000 years. House Sparrows are considered native to parts of Africa, Asia, and Europe. Some scientists worry these birds claim nesting sites that might otherwise go to North American native species, including Bluebirds and Tree Swallows.

Today, House Sparrows can be found on most continents. In many cities, you'll see them nesting on streetlamps and rolling in dust patches to clean their feathers. The dark feather patches on a male's chest are a sign of social status. The bigger the patch, the mightier the male.

LIGHTS OUT IN THE CITY

If you live in a big town or major metropolis, you are probably familiar with light pollution: that's when the skies at night are kept bright by human activity. That pollution can harm other animals. For instance, most bird migration activity happens at night and many species use the light of the moon and stars as guides. When people leave bright lights on, it can confuse birds and cause them to fly off course. Drained, they become easy prey for predators and more likely to fly into glass skyscrapers and other structures (see page 47).

To help wildlife, some communities are participating in "Lights Out" programs. They dim or turn off bright lights overnight. Doing so can save electricity and make it easier for us humans to enjoy the stars too.

KILLDEER

KILLDEER

Charadrius vociferus

APPEARANCE: This long-legged shorebird has white feathers below and brown above. Its large brown-orange eyes are topped with a white "eyebrow" patch and a black forehead band. Adults have two black bands across the chest.

APPROXIMATE LENGTH: 10″ (25 cm)

VOICE: Distinctive "kill-deer" cry

CONSERVATION STATUS: Numbers in decline though still abundant

LOCATION: Common along waterways and open fields in Canada, the United States, and Mexico

DIET: Centipedes, crayfish, snails, and insects

ABOUT

The Killdeer is an abundant shorebird that makes itself at home amidst human activity. They happily nest on beaches, gravel driveways, golf courses, or farmer's fields. Spot them running along and pausing to peck at insect prey. They're also very capable swimmers.

When Killdeer select a nesting site, they perform what's called a "scrape ceremony." As part of this activity, the male and female take turns bowing their chest down and scraping a little hollow in a patch of sand or gravel. Later, the female lays eggs in that spot. Killdeer may scrape several sites to confuse predators that want to find their nest.

MALLARD

FEMALE MALLARDS

Anas platyrhynchos

APPEARANCE: Female Mallard ducks are mottled brown all over with a dark eye stripe and an orange beak. Males, meanwhile, have iridescent emerald heads, yellow beaks, and a thin white collar on their neck. They also have a gray body, brown chest, and black tail. Both males and females have an indigo-blue patch on their wings.

APPROXIMATE LENGTH: 22″ (56 cm)

VOICE: Females "quack" and males have a deep, raspy call

CONSERVATION STATUS: Not a concern

LOCATION: Ponds and lakes across most of North America

DIET: Pondweeds, grasses, and other plants, as well as acorns, crustaceans, insects, and small fish

ABOUT

Across the continent, Mallard families have been spotted visiting swimming pools and exploring backyards. They also boast some awesome aquatic abilities. When they want to feed, they dunk their whole upper body underwater. When it's time to move on, Mallards can launch straight up into the air from the water. Male Mallard ducks are showy but females have a more subdued color scheme that keeps them camouflaged.

MONK PARAKEET

Myiopsitta monachus

APPEARANCE: Vibrant lime green feathers, wings tipped in blue, and a light, off-white chest, chin, and forehead set this colorful parrot apart.

APPROXIMATE LENGTH: 11" (28 cm)

VOICE: Squawks, screeches, and chatter

CONSERVATION STATUS: These South American natives are flourishing in cities across the United States.

LOCATION: Urban areas scattered across the United States are now home to these cheerful, chatty parakeets.

DIET: Fruits, acorns, seeds, and insects

MONK PARAKEET

ABOUT

The Monk Parakeets in North America are descendants of people's escaped pets. These particular members of the parrot family are not tropical birds. They come from temperate grasslands in South America. Today they thrive in cities across the United States with varied climates, including Houston, Chicago, Boston, and San Diego.

One giveaway when these birds are nearby is the presence of one of their massive stick nests. Monk Parakeets build these structures on utility poles, trees, and other elevated locations. They are huge because they serve as home for a whole group of parakeets, a bit like a bird apartment building. One nest could have more than a dozen chambers inside, each for a different pair of nesting parakeets.

ACTIVITY

CAN YOU SPOT THEM ALL?

The birds in this chapter are abundant across North America. If you live in or near a city, you have a good chance of seeing many of them. To practice your birding skills, start a journal to track these sightings. Every time you see one of these species, record the day, time, and location. Make it a game: ask friends if they'd like to compete to see how many species you each see in one month or even one week.

Feeling creative? Adapt your list into a work of art by adding photos or illustrations. You can also stretch your scientific skills by noting the behaviors that you observe. However you customize this birding journal, the goal is to keep a log of what you learn and discover about city birds.

PEREGRINE FALCON

Falco peregrinus

YOUNG, FEMALE PEREGRINE FALCON

APPEARANCE: These Falcons have a dark gray head and back with sideburn patches. Adults have yellow around their eyes and on their beaks and legs. They also have a white throat and black-on-white patterning on their front. Younger birds have gray beak and eye areas, pale yellow feet, and a brown and white pattern on their underside.

APPROXIMATE LENGTH: 17" (43 cm)

VOICE: "Kack-kack-kack"

CONSERVATION STATUS: Their numbers are growing after pesticides in the mid-twentieth century hurt this species.

LOCATION: These birds can increasingly be found throughout the continent, particularly in cities, along coasts, and near the Great Lakes.

DIET: Pigeons, gulls, ducks, and songbirds; small mammals

ABOUT

The Peregrine Falcon is the world's fastest bird. When hunting, they easily clock 70 miles (113 km) per hour. When they dive for prey, they more than double that speed. Found on six continents, these birds may nest on tall buildings or cliff ledges.

RED-TAILED HAWK

Buteo jamaicensis

YOUNG RED-TAILED HAWK

APPEARANCE: This bulky Hawk has a wide, fan-like tail, which, in adults, features cinnamon red feathers.

APPROXIMATE LENGTH: 19" (48 cm)

VOICE: Famous piercing "keeer!" call

CONSERVATION STATUS: Numbers increasing

LOCATION: Across most of North America, except the Arctic

DIET: Small mammals, including rats and rabbits, but also birds and reptiles

ABOUT

Depending on where you live, your local Red-tailed Hawk may sport a very different look. West of the Mississippi River, you may see "dark morphs," with primarily brown-black coloration. To the east are "pale morphs," with white undersides. In the middle, you can find variations in between.

The Red-tailed Hawk's distinctive call is often played on TV shows and movies—even when the camera is showing an Eagle or other large bird with a totally different voice.

ROCK PIGEON

Columba livia

APPEARANCE: Most of these birds have gray heads, iridescent necks, and rounded grayish bodies with black accents. But Rock Pigeons can come in many colors, including all-white, reddish-brown, and richly patterned varieties!

APPROXIMATE LENGTH: 12″ (30 cm)

VOICE: Gentle "cooorr" call

CONSERVATION STATUS: Abundant, common birds

LOCATION: Just about every city or suburb on the continent offers them a good habitat.

DIET: Seeds, grains, and occasional crumbs and other human leftovers

ROCK PIGEON

ABOUT

Many people see pigeons as pests. But Rock Pigeons have been *partners* to humans for a very long time. More than 5,000 years ago, Egyptian hieroglyphics suggest that people were domesticating these birds. We have kept pigeons as pets and livestock. We've asked them to carry messages and race in competitions. In fact, the Rock Pigeons abundant in North America today are descended from tamed birds.

Fun fact:
Rock Pigeons are members of the dove family—in fact, this species used to be called the "Rock Dove."

Humans have prized Rock Pigeons for many reasons. For instance, they are excellent navigators. They can sense the Earth's magnetic field and use visual landmarks and the placement of the sun to find their way home. In groups, these birds also learn about travel routes from one another, sometimes benefiting from generations of knowledge passed through a flock. They can hear low-frequency sounds that our own species cannot. And, with training, Rock Pigeons can master some crazy tricks. Scientists have taught them ping pong, for instance, and how to spot differences between paintings by famous artists. They can even learn how to identify cancer cells on medical imaging.

In the wild, Rock Pigeons are a delight to observe. Be on the lookout for several interesting behaviors, such as bowing and puffing out neck feathers, or dragging their tail along the ground. Both are displays that males use to impress females.

Garden, Park, and Feeder Visitors: Birds and Blooms

AMERICAN GOLDFINCH

Spinus tristis

APPEARANCE: In summer, adult males are sunny yellow with black patches on their heads, wings, and tails. Females are more pale and brown, with no forehead patch. Both have white stripes on their wings and under their tail. In winter, both sexes appear in muted browns.

APPROXIMATE LENGTH: 5" (13 cm)

VOICE: These musical birds have many memorable chirps and songs. Their trills and warbles sometimes sound like "dear-me, see-me" or "po-ta-to chip."

CONSERVATION STATUS: Though common, their numbers are declining in some areas

LOCATION: Spot these finches at feeders and near forests across much of the United States and southern Canada

DIET: Dandelion seeds and thistles are special favorites, though they also eat berries, bark off young trees, and other seeds.

MALE AMERICAN GOLDFINCH

ABOUT

Twice each year, the American Goldfinch transforms. They molt their feathers in dramatic, seasonal costume changes, switching from duller brownish hues in winter to bright yellow and black in summer. By replacing their plumage in this way, these birds are better camouflaged amidst winter's bare tree branches. Then in summer they can look extra attractive when it's time to impress each other.

Bright, sunny-looking males show off their feathers to attract a mate. But female finches also use color cues. Studies have found that female American Goldfinches may recognize each other's social status in part by bill color. Brighter billed females have a higher position in the pecking order.

Being social comes with many advantages. For instance, researchers have found that groups of Goldfinches can hold their own against bigger but solitary species like Northern Mockingbirds, who might otherwise push the little birds away from feeding areas.

FEEDER TIP

Whether you're attracting the American Goldfinches with thistle seeds, Orioles with orange slices, or Gray Catbirds with grape jelly, remember to tidy up bird eating areas and replace food regularly. Clean your feeders at least every two weeks. If birdseed gets wet, swap it out sooner.

ANNA'S HUMMINGBIRD

Calypte anna

APPEARANCE: Glittering fuchsia throat feathers and an iridescent emerald sheen make these little birds truly eye-catching. Adult males have a pink-red crown.

APPROXIMATE LENGTH: 4" (10 cm)

VOICE: Sharp, buzzy "chee-chee-chee"

CONSERVATION STATUS: This bird's range is expanding

LOCATION: Common along the Pacific Coast of North America, from northern Mexico through southern Canada

DIET: Nectar from flowers; they also snatch small insects

MALE ANNA'S HUMMINGBIRD

ABOUT

A group of hummingbirds is sometimes called a "shimmer," and that's a perfect word for this species. For many years, these birds were seen mainly in southern California and Mexico's Baja California peninsula. But thanks in part to human gardeners planting all sorts of interesting flowering trees, Anna's Hummingbirds have expanded their range further north and east.

Like other hummingbirds, Anna's have a distinct body shape. These birds cannot hop or walk. When they fly, their amazing anatomy lets them twist their wings in rapid forward and backward motions so that they hover in midair. When they gather food from a flower, they use their long, forked tongues to reach for nectar through their beak.

ACTIVITY

LEARN HOW TO FEED HUMMINGBIRDS

Want to attract hummingbirds to a green space near you? It's easy to lure these delightful little flyers.

The classic recipe for homemade hummingbird food is straightforward—but you may need an adult's help in the kitchen. Over low heat, dissolve one-part white sugar in four-parts water. Keep things simple: Don't use honey or raw, organic, or brown sugar. Skip the red food coloring and buy a red feeder instead to attract these birds.

Just like any other bird feeder, clean up regularly. If it's hot and sticky outside, swap the water at least twice a week. Rinse out the feeder before you refill it. Never let the water get cloudy.

RUBY-THROATED HUMMINGBIRD

Archilochus colubris

APPEARANCE: Females and young males have glittering green backs and white fronts—males share those features plus a vibrant red throat patch

APPROXIMATE LENGTH: 3″ (8 cm)

VOICE: High-pitched squeaks

CONSERVATION STATUS: Not a concern

LOCATION: Spot these hummingbirds across the eastern half of the United States in the summer. They're also found in much of Mexico, where many of these birds overwinter.

DIET: Flower nectar—they particularly like red and orange tube-shaped flowers, plus occasional insects

MALE RUBY-THROATED HUMMINGBIRD

ABOUT

Bold and beautiful, these feathered jewels enjoy visiting feeders and gardens, where they help pollinate plants. Like other hummers, the Ruby-throated Hummingbird is a high-energy specialist. Sugar-rich nectar helps fuel their active lifestyle. Their heart beats up to 1,200 times each minute and their wings can flap more than 50 times each second.

Given that fast pace, it's equally amazing to see these little birds slow down. As seasons change and temperatures drop, hummingbirds save precious energy by entering a temporary overnight state called "torpor," that's a bit like a mini-hibernation. Their body temperature and heart rate plummet. For several hours, they stay in this cool, energy-saving mode. In the morning, when the sun's warmth returns, their heartbeat and inner temperature climb back up.

PRO TIP

Hummingbirds are territorial. To attract many of these birds, hang multiple feeders or garden with plants they enjoy, such as red bee balm or native honeysuckle.

SIMILAR SPECIES

If you're looking for hummers in the western United States, you may spot a **Black-chinned Hummingbird,** which can look a bit like the Ruby-throated. As their name suggests, adult males of this species have a dark chin, edged in iridescent purple feathers. Females and younger males, meanwhile, have a light white underside and gleaming green feathers above.

BULLOCK'S ORIOLE

Icterus bullockii

APPEARANCE: Adult males are sunny yellow-orange on the front with a black eye stripe, a black throat patch, and another black patch on their head that stretches down their back. Their wings have white accents. Females and young males, meanwhile, have a touch of orange to their face and upper chest but are otherwise gray-brown with white wing patches.

APPROXIMATE LENGTH: 7" (18 cm)

VOICE: Clear whistling song and occasional chattering

CONSERVATION STATUS: Some decline though, overall, not a species of concern

LOCATION: These birds breed across the western United States and overwinter in Mexico

DIET: Fruit, nectar, and insects—they also famously love a freshly sliced orange

MALE BULLOCK'S ORIOLE

Both male and females sing. Their songs may be somewhat different from each other. To human ears, the male's repertoire is a bit more musical, but the female may sing more often.

ABOUT

The Bullock's Oriole is an amazing builder. They can dangle upside down from branches to weave elaborate nests. These architectural wonders require their builders to gather a range of materials, including bits of bark, plant fiber, grasses, and even yarn and hair. The birds then intertwine these strands and strips to craft a hanging pocket structure for laying their eggs.

Often, one parent works on the inside of the nest while the other tackles the outside. The construction can take more than two weeks to complete. By placing their nest on the end of long branches, they make it harder to reach for predators that might want to climb or slither over and steal the Oriole's eggs.

SIMILAR SPECIES

The eastern United States is home to the **Baltimore Oriole**. Male Baltimore Orioles have an all-black head and an intense, sunset orange chest. In the middle of the country, the Baltimore and Bullock's Orioles may meet, mingle, and sometimes have chicks—who grow up to have some features from both parents! Animals whose parents come from two different species are called "hybrids."

BLACK-CAPPED CHICKADEE

Poecile atricapillus

APPEARANCE: A black crown and chin frame white cheeks. This bird's underside is light—with a white chest and pale tan belly, while their wings are gray, edged in white.

APPROXIMATE LENGTH: 5″ (13 cm)

VOICE: Bright "chick-a-dee-dee-dee" call

CONSERVATION STATUS: Populations may be increasing

LOCATION: Forests and feeders from Alaska and much of Canada down through the northern half of the United States

DIET: Spiders, insects, berries, and seeds

BLACK-CAPPED CHICKADEE

ABOUT

These lively little birds have a fascinating social life. Black-capped Chickadees use their "chick-a-dee-dee-dee" call and various songs to communicate with one another. Within a flock, the birds have a chain of command with leaders and followers. Sometimes other species follow Chickadee groups. In fact, other little songbirds, like Titmice and Nuthatches, can recognize Chickadee chatter and will join the band to look for food in winter.

These cute birds are clever and curious. They prepare for cold weather by collecting seeds—including much-loved sunflower seeds from feeders. Black-capped Chickadees then store this food in various hiding places. It's not easy to remember all those secret stashes! But Chickadees are specially equipped for that task. Scientists have discovered that Black-capped Chickadees grow new brain cells every year that may support their memory.

The Chickadee's diet shifts with the passing seasons. They love storing up seeds for the winter but in warmer times, they can collect more than 1,000 insects in a day. They sometimes specialize in gathering certain prey. When feeding their chicks, for instance, Black-capped Chickadees put extra effort into spider hunting, because the arachnids make such nutritious meals.

When it comes to building a nest, Black-capped Chickadees are firm believers in DIY solutions. They prefer to excavate or dig a little space of their own, rather than use a hole that a woodpecker or other bird has left available. Some humans create Chickadee tube nest boxes for these birds that are full of wood shavings so each pair can shape their own little haven.

BLUE JAY

Cyanocitta cristata

APPEARANCE: Distinctive blue crest, back, and wing feathers earn this bird its name

APPROXIMATE LENGTH: 11″ (28 cm)

VOICE: Known for their noisy "jeer" and "jay-jay-jay" call, they also imitate other birds

CONSERVATION STATUS: Not a concern

LOCATION: Common across much of the eastern United States and Canada, they are increasingly found further west.

DIET: They love acorns and peanuts—but will eat just about anything, including plants, nuts, berries, eggs, insects, and even baby birds.

ABOUT

Noisy and bright, the Blue Jay is a favorite for many birdwatchers. They love splashing in a birdbath and their boldness brings drama to your local feeder. They can mimic a Hawk's call to scare songbirds away. They hold their own against bigger animals too. Birders have seen Blue Jays drive off predators like cats and owls. These birds can be inventive. One captive Blue Jay created a paper rake out of newspaper strips to collect seeds.

BLUE JAY

CHIPPING SPARROW

Spizella passerina

APPEARANCE: These little birds have a brick-red cap, a dark stripe across their eyes, and a pattern of light and dark brown streaks over their wings.

APPROXIMATE LENGTH: 5″ (13 cm)

VOICE: Rapid, high-pitched chips

CONSERVATION STATUS: Widespread though declining in some areas

LOCATION: In summer, these birds can be seen across most of North America, including in parks, by feeders, and throughout open areas near forests or human dwellings.

DIET: Beetles, grasshoppers, and other bugs— as well as seeds and grains

ABOUT

Slim and sprightly, the Chipping Sparrow is adept at living around humans. They find most of their food on the ground, diligently seeking out seeds and insects. Chipping Sparrows got their name because—you guessed it—they chip! Researchers have found that, just like human babies babble, Chipping Sparrow chicks make a mix of noises in their earliest days. Over time, they mimic the calls of nearby adults, much like our own species learning to speak.

CHIPPING SPARROW

DARK-EYED JUNCO

Junco hyemalis

APPEARANCE: This stocky little songbird comes in several color patterns. Most have a dark head, a light pinkish beak, and white feathers on the underside of their tails.

APPROXIMATE LENGTH: 6″ (15 cm)

VOICE: High-pitched trill song and occasional chip calls when alarmed

CONSERVATION STATUS: Decline in some areas

LOCATION: These birds breed in Alaska and Canada in the summer and can be found in much of the United States and parts of Mexico in the winter.

DIET: Seeds, berries, and insects

DARK-EYED JUNCO IN THE WESTERN UNITED STATES

ABOUT

Abundant Dark-eyed Juncos flourish in forests and habitats with lots of human presence. You may see them hopping and running along snow or fallen leaves as they look for bugs or visiting your local feeder to gather up seeds. Juncos in some urban and suburban areas have stretched out their breeding season and stopped migrating altogether because there is so much food available in human lawns and gardens. The tradeoff is that living so close to people comes with new dangers, including the presence of feline and canine predators.

Across the continent, Dark-eyed Juncos come in several different colors. East of the Great Plains, these birds are dark slate gray on top and white below.

To the west, they can feature patches of mustard, tan, brown, and rusty red. That variation has made it hard for scientists to determine whether these Juncos are even the same species.

Whatever their colors, Dark-eyed Juncos are welcome winter visitors. In fact, in several parts of the United States, this species is simply called the "snow bird."

HAVE PETS?

There are a few important things to remember about being a responsible animal lover. First, keep cats indoors to protect wildlife and felines. Outdoor cats kill billions of birds each year in the United States and their wandering exposes them to parasites, predators, and passing vehicles. Letting felines explore a catio, or enclosed outdoor space, is a great alternative. If you have dogs, remember that your pet may scare off local wildlife. When you go birding, it can be good to keep Fido at home.

EASTERN PHOEBE

Sayornis phoebe

APPEARANCE: A gray-brown cap and wings adorn an otherwise pale white body. Their beak and legs are dark.

APPROXIMATE LENGTH: 6″ (15 cm)

VOICE: Raspy-voiced "phoebe" call

CONSERVATION STATUS: Not a concern

LOCATION: Eastern Canada and the United States in summer, the southern United States and Mexico in winter

DIET: Millipedes, insects, and berries

EASTERN PHOEBE

ABOUT

The Eastern Phoebe is a member of the Flycatcher family, a group of excellent insect hunters. This solitary bird keeps to itself: even mated pairs seem to prefer their alone time for most of the breeding season. If you spot what could be a Phoebe sitting on a branch, watch its body language. The Eastern Phoebe will pump its tail when a predator is nearby. You can also look for their mud and grass nests, which they often build in nooks on buildings and bridges. In Mexico, the Pacific Coast and the southwestern United States, you may see the **Black Phoebe,** with a soot-dark body and white belly.

GRAY CATBIRD

Dumetella carolinensis

APPEARANCE: A jet black crown, a patch of red-orange under the base of their tail, and an otherwise gray body

APPROXIMATE LENGTH: 9″ (23 cm)

VOICE: Their famous "meow" is just one of several calls they use

CONSERVATION STATUS: Their numbers seem stable

LOCATION: These birds summer over much of the United States—particularly the eastern and midwestern regions. They winter along the Gulf of Mexico.

DIET: Insects and fruit; some birders attract Gray Catbirds with a schmear of dark grape jelly!

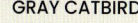

GRAY CATBIRD

ABOUT

Sometimes hearing is believing. Once you've listened to a Gray Catbird's oddly feline call, you won't forget it. These birds produce a range of other noises too. Some sound musical to our ears and others seem a little bit sharp. A Gray Catbird can sing for 10 whole minutes—and that performance has a purpose. Like many songbirds, males croon to proclaim their territory. If you spot their nest, look for their bright turquoise eggs.

HOUSE WREN

Troglodytes aedon

APPEARANCE: Small, brown birds with a long, curved bill and short, pointed tail

APPROXIMATE LENGTH: 5" (13 cm)

VOICE: Their complex songs have a rich mix of burbles and trills.

CONSERVATION STATUS: Not a concern

LOCATION: These little birds can be found from southwestern Canada through the southernmost tip of South America.

DIET: Crickets, moths, beetles, and other insects

HOUSE WREN

ABOUT

The House Wren may not be dazzling to look at, but this little singer's musical talents are spectacular. They are also well known for their flexible attitudes towards real estate. House Wrens build their nest inside of a hollow space called a cavity. They will use an abandoned woodpecker hole, a nook in a building, a broken flowerpot, or even a well-placed boot for their nest site. They may also steal a spot from another species.

Fun fact:
House Wrens sometimes bring spider egg sacs into their nests. When the arachnids hatch, they work as pest control, keeping the area mite-free.

ACTIVITY

BUILDING A BIRDHOUSE

Want to offer nearby nesting birds some housing options? Many species, including Tree Swallows and American Kestrels, will live in a man-made nest box.

You can find a kit and build your own box to help these birds during their breeding season. When selecting a birdhouse, avoid any that include a perch outside the entrance—that feature makes it easier for predators to access the nest.

For Wrens, which are small and comfortable with creative housing options, you can also recycle a familiar household object to create a birdhouse. Some House Wrens will happily nest in a cleaned tin can hung horizontally from a branch. You can also clean out and decorate a milk or juice carton, punching out a small entry hole for the bird, about 1.5" (4 cm) wide.

For any birdhouse, remember to clean out the area at the end of the breeding season, when the chicks have flown from the nest. That prepares the space for another family to move in.

INDIGO BUNTING

Passerina cyanea

MALE INDIGO BUNTING

APPEARANCE: Females are cinnamon brown with lighter, tan bellies. Adult males shimmer in vivid blue.

APPROXIMATE LENGTH: 5″ (13 cm)

VOICE: High clear song phrases, including "What! What! Where? Where? See it! See it!"

CONSERVATION STATUS: Range is expanding though habitat changes and loss may be bringing down their numbers

LOCATION: Across the eastern United States and along the Gulf of Mexico in summer—wintering in Central America and the Caribbean

DIET: Insects, spiders, seeds, and plants

ABOUT

Indigo Bunting males perch on high branches or atop powerlines to sing. They learn these tunes from other buntings nearby. As a result, these birds have "song neighborhoods." Females, meanwhile, are builders: They construct cup-shaped nests with plant bits, such as grasses and strips of bark. Indigo Buntings migrate at night and navigate by looking at the stars. In the western United States, you might notice the **Lazuli Bunting**. Like the Indigo, Lazuli females are brown while males sport bright blue feathers plus white bellies edged in orange at their chest.

PAINTED BUNTING

Passerina ciris

FEMALE PAINTED BUNTING

APPEARANCE: Females and young males are vibrant lime green on top and lemon-yellow below. Adult males have yellow-green backs, bright crimson bellies, and blue heads.

APPROXIMATE LENGTH: 5″ (13 cm)

VOICE: Singsong whistles and warbles

CONSERVATION STATUS: Numbers may be declining in part because people illegally capture these birds for the pet trade

LOCATION: Throughout forests and garden edges in the southeastern United States (particularly along the Gulf), as well as Mexico and the Caribbean

DIET: Seeds and insects

ABOUT

Female Painted Buntings build nests and feed chicks while males perch high, sing loud, and defend their territory fiercely. During the breeding season, when males want to get a mate's attention, they spread out their gloriously colored feathers like miniature turkeys on display.

MALE PAINTED BUNTING

MOURNING DOVE

MOURNING DOVE

Zenaida macroura

APPEARANCE: A small head and broad body with an elegant long tail—they have a peachy-brown underside and light gray-brown wings with black accents above

APPROXIMATE LENGTH: 11" (28 cm)

VOICE: Lovely, haunting "coo-ah, coo, coo, coo"

CONSERVATION STATUS: Not a concern—this bird's numbers may even be increasing

LOCATION: In forests, farms, towns, cities, and suburbs across much of southern Canada, the United States, and Mexico

DIET: Seeds from grasses and other plants

ABOUT

Before you see these doves, you'll hear them. Their sweet cooing call is sometimes confused with an owl's hoot. They can also make a rustling "swoosh" sound with their wings, which they may use to warn one another when danger is nearby. These birds are adaptable—meaning they can live in many different habitats—and abundant. One pair of Mourning Dove parents may hatch eggs and raise chicks six times in a single year.

NORTHERN CARDINAL

Cardinalis cardinalis

APPEARANCE: Females are a soft brown with pops of red, including on their crest and beak. Males are scarlet all over. Both have a darker black face mask. Young Cardinals of both sexes have a dark beak that becomes red as they mature.

APPROXIMATE LENGTH: 8.5" (22 cm)

VOICE: Loud, repetitive whistles, sometimes singing the phrase "cheerily, cheer, cheer, cheer"

CONSERVATION STATUS: Not a concern

LOCATION: Spotted in yards, gardens, towns, and even near deserts throughout the eastern United States, some southwestern states, and parts of Mexico

DIET: Seeds, fruit, and insects

ABOUT

FEMALE NORTHERN CARDINAL (RIGHT AND MALE NORTHERN CARDINAL (LEFT

These crested crimson flyers have a ton of style. The Northern Cardinal is easy to spot and rewards careful observation. Notice, for example, how their crest can be raised—to signal agitation or excitement—or lowered when they are relaxed. Cardinals of both sexes sing. Females may be singing from the nest, for instance, to tell their mates to bring food. Males, meanwhile, use their tunes to proclaim their territory.

NORTHERN MOCKINGBIRD

Mimus polyglottos

APPEARANCE: Gray above, dark wings with white wing bars, and white feathers below; this bird has light brown eyes, a dark eye stripe and long, slim tail feathers.

APPROXIMATE LENGTH: 10" (25 cm)

VOICE: Can imitate dozens of sounds, including House Wren songs, dog barks, toad croaks, and car alarms

CONSERVATION STATUS: Their numbers have rebounded since the end of the pet trade, in the early 1900s, when this tuneful bird was captured and caged so people could listen to their music.

LOCATION: Found across most of the United States and Mexico

DIET: Insects, fruits, and berries

NORTHERN MOCKINGBIRD

ABOUT

Ever wanted to hear nature sounds on shuffle? The Northern Mockingbird can help. This brilliant mimic sings day and night. They may cycle through about 200 different songs and copy varied sounds from their environment.

These birds are very protective and will dive bomb you or your pet if you get too close to their home. Northern Mockingbirds sometimes nest in multiflora rose bushes, which offer dense, thorny shelters that are perfect for raising chicks. As a bonus, the birds can nibble on rosehip berries. Mockingbirds have a distinctive wing flash display, in which they abruptly spread wide their wings, showing off white patches across them. Some scientists think the birds do this to startle insect prey.

ACTIVITY

BIRDSONG BASICS

Northern Mockingbirds master their music through careful listening and imitation. Birders can do something similar. We have very different mouths and vocal chords, so we can't reproduce the sounds birds make. But we can use a system called "mnemonics," or little phrases that remind us of a bird's call.

This book has many mnemonic examples, including the call of the Black-capped Chickadee and Eastern Phoebe (who seem to announce their own names) and the repeated phrases of American Goldfinches and Indigo Buntings. You can hear many of these calls online through sites such as Audubon.org/Bird-Guide and AllAboutBirds.org.

For fun, write your own mnemonic lyrics as you listen!

BIRD-FRIENDLY CONTAINER GARDENING

One of the best ways to make birds feel welcome is by offering them a little patch of green. But you don't need a yard to start gardening. In fact, with the right plants, you can set up pots on a windowsill or balcony to benefit birds.

The best garden options will depend on where you live, the amount of light and space available, and the flowers and shrubs native to your region.

For inspiration, here are a few plant varieties that thrive in containers and attract birdlife:

Blueberries: These shrubs may need a few years to offer fruit, but they're worth the wait. The berries are a treat for Warblers, Waxwings—and humans!

Columbine: This flowering plant requires a 12-inch (30 cm) wide pot and will bloom every spring, much to the delight of hummingbirds.

Coneflower: A deep pot can be perfect for many varieties of coneflower, which produce colorful flowers and, later, seeds for Northern Cardinals, Blue Jays, American Goldfinches, and other birds to enjoy.

Goldenrod: This easy-to-grow plant will spread and fill your container with yellow blooms in the late summer and early autumn. Tufted Titmice, Black-capped Chickadees, and Dark-eyed Juncos will collect the seeds later.

Milkweed: Many people grow this flowering plant for Monarch butterflies, which use it to lay their eggs. But Orioles and Goldfinches also love milkweed, using it to build their nests.

Sunflowers: Pick a small variety, set it in a bright spot and enjoy its cheerful blooms! When the flower goes to seed, finches, Cardinals, and other birds will come and feast.

ROSE-BREASTED GROSBEAK

Pheucticus ludovicianus

APPEARANCE: Females and young males are light yellow below and mottled brown above with a distinctive white eyebrow. Adult males, meanwhile, look like they're dressed for the opera—with the voice to match! They have a formal-looking pattern with a black head and back feathers, white underside, and striking crimson chest.

APPROXIMATE LENGTH: 8″ (20 cm)

VOICE: Lovely, delicate tune that sounds a bit like an American Robin

CONSERVATION STATUS: May be losing habitat

LOCATION: In summer, primarily found in or near forests of the eastern United States, also much of southeastern Canada and along the Gulf Coast of Mexico; winters in the Caribbean and Central and South America

DIET: Seeds, berries, and insects, including caterpillars and beetles

MALE ROSE-BREASTED GROSBEAK

Despite the male's striking good looks, you're more likely to hear this bird than see it. They conceal themselves well in foliage and nest high up in trees. That said, you may spot them at a feeder, using their large bills to crack open raw peanuts and safflower seeds.

ABOUT

The songs of the Rose-breasted Grosbeak are a joy for the ear: beautiful melodies that string together rising and falling notes. Both male and female grosbeaks sing. In fact, mated pairs will croon softly to each other when they take turns sitting on their eggs.

Rose-breasted Grosbeaks make good parents. Male and female each spend several hours sitting on their eggs. They can also be fierce when it comes to protecting their home. Pairs will gang up on predators to scare them away and males will dive at other unwelcome males who dare to sing in their territory.

SIMILAR SPECIES

Birders in the western United States get to enjoy the sweet musical stylings of the **Black-headed Grosbeak.** While female Black-headed Grosbeaks and Rose-breasted Grosbeaks may look alike, the males are more distinct. The Black-headed Grosbeak has—no surprise—a black head and largely black wings. Their neck and lower body are orange, flecked with yellow. In the Great Plains of North America, the Black-headed and Rose-breasted Grosbeaks interbreed. Their chicks can have patches of pink, black, and orange.

SONG SPARROW

Melospiza melodia

APPEARANCE: Covered in white-gray and chestnut brown stripes—their coloration varies in different parts of the continent

APPROXIMATE LENGTH: 5" (13 cm)

VOICE: These birds have a rich range of songs, including one that sounds a bit like "Madge-Madge-Madge, put-on-your-tea-kettle-ettle-ettle"

CONSERVATION STATUS: Abundant in many areas

LOCATION: From Alaska and Canada through to northern Mexico, this bird is common in gardens and marshlands.

DIET: Beetles, ants, and other insects—as well as seeds from grasses and weeds

SONG SPARROW

ABOUT

These tuneful songbirds can be hard to spot. The Song Sparrow is often well hidden in dense shrubbery and vegetation. But if you catch their cheery, rich songs and wait patiently, you may observe this little bird bursting from a rose bush or hedge. Song Sparrows are fiercely territorial. If a new bird gets too close, a defending sparrow will copy the newbie's song, wave a wing menacingly and—if necessary—attack the incoming bird.

Fun fact:
Song Sparrows are paler in desert regions but darker in cooler climates. In central Mexico, some of these birds have white throats and black streaks.

LITTLE BROWN BIRDS: SPARROWS VS FINCHES

The difference between a sparrow and a finch isn't always obvious. Even expert birders will occasionally lump species together as "little brown birds" or "little brown jobs!"

Both sparrows and finches are small songbirds with chunky bills that help them chomp down on seeds. Many sparrow and finch species come in brown or brown-streaked feathers that offer perfect camouflage in dense branches.

But to the observant eye, there are a few things that set these birds apart. Finches have shorter legs than sparrows and often perch with their tails angled down. In addition, you're more likely to notice finches on top of trees and shrubs, while sparrows keep closer to the ground.

TREE SWALLOW

Tachycineta bicolor

APPEARANCE: Iridescent blue covers this bird's head and back, complemented by a snowy white underside. Tree Swallows have long wings that arc to form a distinctive silhouette when flying. Compared with Barn Swallows (page 75), they have a stubbier, squarer tail.

APPROXIMATE LENGTH: 5" (13 cm)

VOICE: Burbling chirps and tweets

CONSERVATION STATUS: Common through their range but declining in numbers

LOCATION: Abundant across most of the continent, you may notice them swooping and gliding over lakes and ponds

DIET: Mosquitoes, flies, and other insects caught mid-flight make up most of their meals—though they will also eat bayberries and occasional seeds

TREE SWALLOW

ABOUT

Swallows and Swifts are often called flying acrobats. The Tree Swallow is no exception. As evening approaches, these birds zip past in large groups, swirling and twisting in a cloud of activity. If you are near a patch of freshwater, you may catch similar performances throughout the day.

All of that twirling has a purpose. This bird is perfectly designed to pursue flying insects. In a single day, an adult Tree Swallow may snag hundreds of bugs. When raising chicks, Tree Swallow parents will collect multiple insects on a single hunting trip, gathering the bunch into a ball in their beak.

Because of their hunting prowess, some people invite Tree Swallows to their farms and gardens by setting out nest boxes for these birds. The Swallows offer a chemical-free solution for pest control—with a beautiful flying performance to boot!

Fun fact:
Tree Swallows are playful. One bird will catch a feather in the air, and another will chase after. When the first drops the feather, the second will swoop in to catch it.

SIMILAR SPECIES
Throughout western Canada, the United States and Mexico, the **Violet-green Swallow's** gleaming emerald back, purple rump and white front make for a striking birdhouse visitor.

TUFTED TITMOUSE

Baeolophus bicolor

APPEARANCE: Little birds with a silvery crest and back and a white front; they have a small patch of orange feathers below their wings

APPROXIMATE LENGTH: 6″ (15 cm)

VOICE: Whistled "peter-peter-peter" song

CONSERVATION STATUS: Not a concern—some scientists think that bird feeders have helped this little bird thrive across an even larger area than in the past.

LOCATION: Common in much of the eastern United States, particularly in areas with deciduous trees

DIET: Seeds—they love sunflower seeds—as well as nuts, berries, and insects

TUFTED TITMOUSE

ABOUT

In many parts of the United States, this dainty crested bird is a familiar feeder visitor. Look for Tufted Titmice hopping along branches or hanging upside down as they gather insects and seeds. Often, they carry their food a little ways away to sort and store it for later.

To crack acorns and other hard snacks, the Tufted Titmouse will hold a seed steady with their feet and pound it with their bill. These birds are fairly solitary. They do not gather in flocks but instead each keep and protect a tight territory. That said, they are family minded. Often, one chick will stick around as a helper for its parents, babysitting younger siblings. In Mexico and Texas this bird has a closely related cousin, the **Black-crested Titmouse.**

ACTIVITY

CREATE A PINE CONE FEEDER

Offer local birds a nutritious bounty of seeds with a simple pine cone bird feeder. Spread a layer of peanut butter on a large pine cone and roll it in black oil sunflower seeds (or the birdseed mix of your choice). Tie some string to the feeder and you can hang it from a balcony or tree branch!

Depending on where you live, you could easily draw Chickadees, Titmice, and even certain woodpeckers. Setting this protein-rich feeder out in winter, when less food is available, will draw especially eager visitors.

WESTERN BLUEBIRD

Sialia mexicana

APPEARANCE: Adult males have bright blue heads and wings, with a rust-colored midsection and lighter feathers on the lower white underparts. Females have a subtler, grayer variation of this color palette, with hints of orange and blue.

APPROXIMATE LENGTH: 7" (18 cm)

VOICE: "Cheer-lee, churr" song

CONSERVATION STATUS: Loss of forest habitat has hurt this bird, though people are taking steps to support the species.

LOCATION: Within select stretches of the western United States and parts of Mexico, particularly in open areas along the edges of woodlands

DIET: Mainly insects—including grasshoppers, caterpillars, and beetles—though they also enjoy juniper and mistletoe berries and other small fruit

MALE WESTERN BLUEBIRD

FEMALE WESTERN BLUEBIRD

ABOUT

Among Western Bluebirds, bird families may live in little "kin neighborhoods," sharing a rich territory. During the breeding season, many Western Bluebird parents receive help from a younger female bird who is not one of their own offspring. Often this young Bluebird is new to the area. The group sticks together through the winter, sharing berries and warmth. They even defend the nest together. In the spring, the young female helper may pair off with one of the sons from the local nest—a bit like "marrying into the family."

Putting up a Bluebird nest box is a great way to help Bluebirds near you and an awesome way to observe these colorful birds more often.

SIMILAR SPECIES

The **Eastern Bluebird**—found in the eastern United States—looks similar to the western variety. In between these two species is the **Mountain Bluebird** who lives along the Rocky Mountain Range. This species has the most distinct appearance of the three: Adult females are brown and gray with just a hint of blue on their wings and tails. Adult males are striking in sky-blue feathers with just a touch of white below.

WHITE-BREASTED NUTHATCH

Sitta carolinensis

APPEARANCE: A sharp, upturned beak gives this and other Nuthatches a distinctly different profile from other small birds. Distinguishing features include a dark cap, gray-and-black-tipped wings, and a largely white underside

APPROXIMATE LENGTH: 5.5" (14 cm)

VOICE: Loud "yank-yank-yank" call

CONSERVATION STATUS: Numbers are increasing

LOCATION: Forests and clearings across most of the United States and parts of central Mexico

DIET: Insects and spiders throughout the summer and seeds in the winter

WHITE-BREASTED NUTHATCH

ABOUT

In warmer months, these little woodland residents can be spotted hopping up and down trees, hunting for insects. They use their angled beak to chip away bark to find prey. In wintertime, however, they are also common at feeders where they enjoy protein-rich peanut butter and suet.

Nuthatches sometimes smear insects around the entrance to their nest area—often a little hole or cavity in a tree. It's possible this activity releases a smell that keeps predators away.

You may see a White-breasted Nuthatch spread its wings wide and sway. This behavior may be intended to scare off other birds and squirrels.

ACTIVITY

DIY BIRDBATH

Whether taking a sip or an all-out splash-session, many birds love a good bath. Access to clean freshwater is important for hydration and for keeping a bird's feathers clean. Scientists have found that if European Starlings can't bathe regularly, for instance, their flight becomes clumsier.

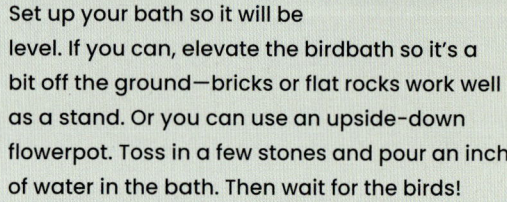

Making your own birdbath is easy. You can repurpose a flowerpot tray or a shallow baking tin. Set up your bath so it will be level. If you can, elevate the birdbath so it's a bit off the ground—bricks or flat rocks work well as a stand. Or you can use an upside-down flowerpot. Toss in a few stones and pour an inch of water in the bath. Then wait for the birds!

Just like a birdhouse or feeder, maintenance is important. Change out the water weekly and rinse the bath with vinegar between uses.

WHITE-THROATED SPARROW

Zonotrichia albicollis

APPEARANCE: This sparrow comes in two varieties, with either tan or white striping above the eye. Both have a buttercup yellow patch by their beak, white chins, black eye stripes, and wings streaked in brown and black.

APPROXIMATE LENGTH: 7" (18 cm)

VOICE: "Oh-sweet-Canada-Canada-Canada"

CONSERVATION STATUS: Overall, not a concern—though their numbers are in decline in certain areas

LOCATION: Much of Canada and the eastern United States as well as the Pacific Coast

DIET: Seeds and insects

WHITE-THROATED SPARROW WITH WHITE STRIPED COLORATION

ABOUT

This little sparrow has an unusual approach to pairing off: within this one species there are two types of male and two types of female White-throated Sparrow. With a little observation, you may notice that some of these birds have white stripes above the eye while others have tan stripes. It turns out that white-striped males generally mate with tan-striped females—and tan-striped males with white-striped females. These color-coded sets also behave in different ways. The white-striped group tends to be louder and more aggressive while, tan-striped sparrows bring chicks more food.

Birdsongs change over time and across territories. For instance, scientists have noticed a new "Oh-sweet-Cana-Cana-Cana" tune catching on in this species.

ACTIVITY

BIRD SAFE WINDOWS

In the United States alone, up to one billion birds die each year because they collide with buildings. Windows are especially dangerous. Highly reflective glass acts like a mirror, reflecting surrounding greenery and confusing a bird into thinking it has found a safe path to fly. Collisions happen day and night (see page 22).

Several steps can make windows safer. For example, keep blinds partially closed or ask your parents to set up window screens. Another solution: stick something on the outside of windows to help birds see the glass. You can buy special stickers for this purpose or create your own by adding paper shapes or patterns to decorate the window in densely spaced patterns. Tape them on the outside glass and space them about 2 to 4 inches (5–10 cm) apart. Repeat your design across the pane. A single cut-out may be pretty, but it won't be enough to warn incoming birds. A checkerboard grid, on the other hand, is simple and effective.

3

Lake, River, and Marsh Birds: Freshwater Fans

BALD EAGLE

Haliaeetus leucocephalus

APPEARANCE: Adult Bald Eagles have white heads and tails and huge dark chocolate-brown bodies. They also have yellow beaks, eyes, legs, and feet. When they're very young, Bald Eagles are often muddy brown all over and, as they get older, have more and more white feathers.

APPROXIMATE LENGTH: 33″ (84 cm)

VOICE: Squeaky and high-pitched, sometimes sounds like a giggle

CONSERVATION STATUS: About 50 years ago, a combination of pesticides, pollution, habitat loss, and hunting drove their species nearly to extinction. Today, however, their numbers have bounced back in many places.

LOCATION: Along forested waterways throughout North America

DIET: Mostly fish and shellfish, but also the occasional bird and small mammal

BALD EAGLE

ABOUT

This American icon has some serious skills. For instance, Bald Eagles and other raptors can see details at 20 feet (6 m) that most humans can't see until they're 5 feet (1.5 m) away. That's where we get the expression "eagle-eyed!"

Bald Eagles are good fishers and can even wade into shallow water for a meal. Birders have captured videos of Bald Eagles swimming to drag their catch along in the water. Even with these awesome abilities, Bald Eagles are not above scavenging someone else's leftovers or stealing food. They will nab fish caught by Ospreys, for example, rather than find their own dinner.

Bald Eagles also have a fascinating personal life. They hunt on their own but mate for life, rebuilding their stick nests with their mate each year in tall trees. Those nests are often enormous: more than 4 feet (1.2 m) wide and 2 feet (.6 m) deep. Like several other raptor species, these birds have some spectacular behaviors around choosing a mate. In a "cartwheel courtship flight," a pair of Eagles fly high into the sky, lock talons, and spin around each other like swing dancers as they descend. Talk about falling for each other!

BELTED KINGFISHER

Megaceryle alcyon

APPEARANCE: Females have a band of rust-brown feathers across their chest. Both sexes have a prominent head, with a crest of blue-gray feathers and long, pointed beaks—their backs are also slate blue.

APPROXIMATE LENGTH: 12" (30 cm)

VOICE: Rattling voice

CONSERVATION STATUS: May be in decline

LOCATION: Lakes, ponds, rivers, and streams across most of the continent; many of these birds breed in Alaska and Canada and winter in the southern United States and Mexico.

DIET: Crayfish, small fish, and other aquatic life such as frogs and tadpoles

FEMALE BELTED KINGFISHER

ABOUT

As their name suggests, there's something regal about this brightly colored, crested bird. The Belted Kingfisher lives along the water's edge and relies on a very specific habitat to raise its young. These birds seek out steep banks of soft sand or dirt. They spend several days excavating a tunnel in those conditions, creating a burrow that can be 6-feet (1.8 m) deep for laying their eggs.

Kingfishers need easy access to water for hunting. They also want their eggs to be difficult for predators to reach. Those nesting needs limit the places where these birds can safely raise a family. That means when humans change waterways, for instance, by building dams, it can have serious consequences for these birds.

Like many freshwater and coastal birds, the Belted Kingfisher has a special technique for catching its dinner. These birds hover over the water—flapping their wings while keeping their head still to watch for prey—and then dive headfirst with their eyes closed. Once they've got a catch, they smack their wriggling meal against a rock or branch before tossing it into the air, catching it with their beak, and swallowing their prize whole.

Fun fact:
Belted Kingfishers have been in North America for a long time. Some of their fossils date back 600,000 years!

CANADA GOOSE

Branta canadensis

APPEARANCE: A large bird with a long black neck, white chin strap, brown-gray upper body, and light belly. There are several varieties (or "subspecies") of Canada Goose, with smaller geese in northern regions and darker-colored geese in the west.

APPROXIMATE LENGTH: 35" (89 cm)

VOICE: "Honk!"

CONSERVATION STATUS: Thanks to conservation work, these once declining birds are now abundant.

LOCATION: Many of these geese breed across Alaska and Canada in the summer and migrate to the southern United States in the winter. Others stick around all year in the regions between.

DIET: Grasses, grains, seeds, and berries

ABOUT

This elegant migratory bird is a symbol of the changing seasons in many parts of the continent. Males and females look alike. In fact, they seem to prefer mates of a similar size. But they behave in ways that can help you tell them apart. While females may sit on eggs and stay closer to chicks, males are often standing guard, keeping an eye on any possible predator. If you see a Canada Goose pumping its head, opening its bill, honking, or hissing, keep your distance! It may be warning you away from its family.

CANADA GOOSE

MIGRATION MASTERS: WHY AND HOW BIRDS TRAVEL

Many animals—including fish, butterflies, and deer—journey every year to access food, water, and other essentials. In truth, how migration first began is a bit mysterious, but it's clear that when animals embrace long-distance trips, they can enjoy certain advantages. For instance, some birds find a safe place with lots of food in summertime for raising their young, then travel south to escape wintry weather.

Migratory species are born with many abilities to help them travel. Some, for instance, can read the stars like a map or sense the Earth's magnetic field to identify North and South. It's also clear that experience matters. When Canada Geese migrate, for instance, the older birds are often near the front of their V-shaped groups, leading the way.

COMMON LOON

Gavia immer

APPEARANCE: Adults change color in the seasons: in winter, these loons are gray above and white below. In summer, they have a beautiful pattern of black and white stripes and spots that set off their garnet-red eyes.

APPROXIMATE LENGTH: 30" (76 cm)

VOICE: Haunting calls include an eerie laugh and yodel

CONSERVATION STATUS: These birds need clean, freshwater habitats that may be at risk due to climate change and pollution.

LOCATION: Both the east and west coastlines of North America and, in summer, lakes throughout Canada and the northern United States

DIET: Minnows, perch, cod, and other fish

COMMON LOON IN SUMMERTIME

ABOUT

On a calm night, the call of the Common Loon echoing across the water is unforgettable. Their wails help pairs find one another. In summer, males yodel to stake out their territory.

The Common Loon spends almost all of its time swimming and paddling. They waddle ashore only to mate and sit on their eggs. Whether on freshwater or seacoasts, these birds are perfectly adapted to an aquatic existence. Their body shape helps them dive much more quickly than a human. They shoot below the waves like little torpedoes. Their bones are solid and heavy, unlike most birds, which helps them sink.

Though most of their dives are quick, researchers believe the Common Loon can delve some 200 feet (61 m) below the water's surface and stay underwater for 5 minutes.

These birds hunt by sight and rely on clean, clear water. When lakes become polluted, loons cannot raise their young there. Some people, who call themselves "Loon Rangers," work to educate people about lake protection and conservation to help Common Loons.

Fun fact:
Baby Loons often hitch a ride on a parent's back. Doing so helps little ones keep warm, take swim breaks, and avoid predators.

COMMON YELLOWTHROAT

Geothlypis trichas

APPEARANCE: These little birds have cheery, butter-yellow throats and bellies, topped with an otherwise greenish-brown body. Males have a black face mask edged in white.

APPROXIMATE LENGTH: 5" (13 cm)

VOICE: Fast "witchety-witchety-witchety" song

CONSERVATION STATUS: Their numbers are falling, although they are still a widespread species.

LOCATION: Swamps, wetlands, marshes, and upland areas in much of Canada, the United States, and Mexico

DIET: Dragonflies, mayflies, and other insects

ABOUT

MALE COMMON YELLOWTHROAT

Who is that mysterious masked warbler? If you spotted him in a scrubby wetland habitat, he is likely a Common Yellowthroat.

In the midwestern United States, these birds are sometimes called "yellow bandits." Adding to their air of mystery, they keep to thick vegetation, like dense, low-growing shrubs and grasses. These birds nest on or near the ground and so Common Yellowthroat parents make a point of approaching their chicks from one direction and leaving in another to better conceal their nest.

DOUBLE-CRESTED CORMORANT

Nannopterum auritum

APPEARANCE: Large, lean, and long-necked, with dark, mostly black coloration and orange accents around their beak and eyes

APPROXIMATE LENGTH: 30" (76 cm)

VOICE: Throaty grunts that sound a bit like oinking pigs

CONSERVATION STATUS: Their numbers may be on the rise in some areas

LOCATION: From coast to coast, across aquatic habitats

DIET: Fish, crabs, frogs, and other food harvested from the water

ABOUT

DOUBLE-CRESTED CORMORANT

The dramatic Double-crested Cormorant is often spotted with its wings wide, soaking up rays. Unlike many other coastal and freshwater birds, the Cormorant's feathers have a structure that holds onto water droplets. That difference may help them dive, but it also means they need some sunshine after they surface to dry out. These social birds live in groups, called colonies. During the breeding season, the inside of the Double-crested Cormorant's mouth is bright blue!

GREAT BLUE HERON

Ardea herodias

APPEARANCE: This large, leggy bird has a six-foot (1.8 m) wingspan as an adult. It has a long yellow beak and legs, its coloration is bluish-gray, with brown thighs and a white head. During the breeding season, these Herons have a striking dark stripe over their eyes with long feathers trailing down their neck. Their eyes are yellow.

APPROXIMATE LENGTH: 45″ (1.1 m)

VOICE: Harsh "frahnk" squawk

CONSERVATION STATUS: Not a species of concern

LOCATION: From southern Alaska, along the Pacific Coast, throughout much of the United States and Mexico—look for them flying over wetlands or wading into fresh or salt water

DIET: Fish, frogs, snakes, rodents, insects, and birds

GREAT BLUE HERON

ABOUT

Statuesque and still, the Great Blue Heron is a master of patience. This common visitor to marshes, swamps, rivers, and lakes has a varied diet. These birds will wait at the water's edge for a long time for their next meal. When they do move, it's either very slowly—to avoid scaring a skittish fish or frog—or lightning fast.

When they do strike, they may stab with their sharp beaks. Birders have also spotted the Heron gulping down snakes like spaghetti. They even catch and eat small mammals, such as gophers. The bones in their neck give them flexibility to uncoil their neck and snap at prey. When they're not hunting, however, they can tuck their neck in tight beside their body, creating an S-shape.

Watching these Herons in action, it's not so hard to imagine the connection between today's birds and the ferocious dinosaurs of the past (see page 70).

The Great Blue Heron is a fascinating bird in many respects. Despite their size, these birds only weigh about 5 pounds (2.3 kg), thanks to their hollow bones. Great Blue Herons can swim, wade, wander on land, and fly with grace. They also have excellent night vision.

Fun fact:
An all-white variety—
the Great White Heron—
lives in Florida and
the Caribbean.

BIRD BEAK EXPERIMENT

As you page through this chapter, you may notice that many freshwater and coastal birds have distinctive beaks. Why *is* that? Try this at-home experiment to find out.

Start by gathering objects that work like a bird's bill. For example, Pelicans and ducks strain food from water through their beaks, so grab a sieve or slotted spoon to represent those birds. Some shorebirds, such as the Wilson's Snipe, have long thin beaks: tweezers or chopsticks are a good stand in. Herons and Egrets, meanwhile, pick up prey with much longer beaks, so you can look for kitchen tongs.

Next, it's time to put your beaks to the test. Create two environments for catching "prey" with two bowls. In one, fill the bowl with water and add marbles or dominoes to practice grabbing slippery fish. In the other, add cut up rubber bands—a bit like worms or little crustaceans—and pour in sand. (If you don't have sand, dirt or even breadcrumbs will work.)

Now try out different beaks in the bowls. Do some work better than others at capturing prey? What advantages do you notice for each?

As you experiment, pay attention to how different strategies suit different environments. The amazing variation in beaks and bills has allowed birds that live along the water to specialize in how they find and eat their food.

THE WILSON'S SNIPE IS A SMALL SHOREBIRD WITH A DISTINCTIVE LONG, THIN BEAK

OSPREY

Pandion haliaetus

APPEARANCE: Long-legged raptors with white bellies and faces, mostly brown wings and a black eye stripe

APPROXIMATE LENGTH: 21" (53 cm)

VOICE: Excited chirp "twer-twer-twer!"

CONSERVATION STATUS: Once imperiled by pesticides, these birds have made a comeback.

LOCATION: These birds soar above rivers, lakes, and coasts throughout North America. Many migrate from Canada and the United States to Mexico and Central America for the winter.

DIET: Flounder, shad, and other fish

OSPREY

ABOUT

Found on every continent except Antarctica, Ospreys have many adaptations that make them master fishers. For instance, they have small barbs on the bottom of their feet and can arrange their toes to clamp around prey. Their fishing technique is also impressive. These birds will perch with a view of the water. When they see something promising, they hover over the waves to watch before shifting their wings into a sharp V-shape and diving feet first. They're also good travelers. One bird may fly more than 160,000 miles (257,495 km) in a lifetime of migrations.

RED-WINGED BLACKBIRD

Agelaius phoeniceus

APPEARANCE: Adult males are totally black except for their bright scarlet shoulder patch, edged in yellow. Adult females, meanwhile, are streaked in brown and white, with a very faint blush of red on their shoulders.

APPROXIMATE LENGTH: 8" (20 cm)

VOICE: Raspy "o-ka-leeee"

CONSERVATION STATUS: Though widespread, their numbers are declining.

LOCATION: Marshes, swamps, and fields near freshwater across most of the continent except Arctic regions

DIET: Seeds, grains, insects, and berries

ABOUT

The bold Red-winged Blackbird is fierce and territorial. You may spot the flash of a male's red feather patch as he flaps into action, chasing out other males and intruder birds. In fact, these birds may spend a quarter of each day on the defensive. Why so protective? In this species, one male may have several female mates, all nesting on his territory.

MALE RED-WINGED BLACKBIRD

When they are not in breeding season, these birds mellow quite a bit. They relax their aggressive attitudes and gather together in big communal groups overnight, called roosts.

SANDHILL CRANE

Antigone canadensis

APPEARANCE: This large, mostly gray bird has a bright red patch over their eyes, framed by white cheeks. These birds have long, dark bills and long, curved feathers that fall over their tails.

APPROXIMATE LENGTH: 47" (119 cm)

VOICE: Distinctive, rattling "kar-r-r-r-o-o-o"

CONSERVATION STATUS: Their range is expanding, though some habitat is in jeopardy. Today, the Sandhill Crane is one of the few Crane species in the world that is neither endangered nor threatened.

LOCATION: Many of these birds nest and breed in Alaska and northern Canada, then travel south to winter in Mexico and the southern United States.

DIET: Plants, insects, rodents, reptiles, amphibians, berries, grains, and even young birds

SANDHILL CRANE

ABOUT

Each spring, Sandhill Cranes perform dazzling dances. They gather in open fields and grasslands, trumpet loudly, and flap their wings. They bow, hop, run, and leap. This stunning choreography is used to impress mates. Once they pick a partner, they may stick together for more than a decade, dancing for years to come.

MALE WOOD DUCK

WOOD DUCK

Aix sponsa

APPEARANCE: This duck species has a distinct feathery cap. Females are mottled in gray, brown, and white feathers. Look for the female's white, comet-shaped eye patch. Adult males shimmer with iridescent green on their head, chestnut brown on their chest, and many striking patches, including black and white along their neck and wings.

APPROXIMATE LENGTH: 20" (51 cm)

VOICE: Females exclaim "woo-eek!" while males have a softer "ter-weee?" call.

CONSERVATION STATUS: Hunting and loss of nesting sites hurt this species for a long time, but their numbers may be recovering.

FEMALE WOOD DUCK

LOCATION: Found in forested wetlands throughout much of the eastern United States and some of the Pacific Coast—their range may be expanding

DIET: Plants, fruits, and nuts

ABOUT

The colorful male Wood Duck keeps in tip-top health to impress females. Females, meanwhile, handle parenting independently. They nest in trees, often far from the water's edge, using their strong claws to climb up bark.

4

Wetland and Coastal Birds: Waders and Beachgoers

BROWN PELICAN

Pelecanus occidentalis

APPEARANCE: This massive seabird has a distinctive bill, equipped with an expandable pouch. Young Pelicans are pale below and gray-brown above. Adults are brownish gray all over with pale yellow heads. Adults have dark brown necks during the breeding season. If you are looking up at these birds from below, the lighter bellied birds are young Brown Pelicans and the darker ones are more mature.

APPROXIMATE LENGTH: 51" (1.3 m)

VOICE: Silent except for occasional grunts

CONSERVATION STATUS: In the 1950s, widespread use of a pesticide called DDT for insect control started to harm many animals. In Pelicans, this chemical weakened bird eggs, leaving them extremely fragile. Happily, their numbers have rebounded since people stopped using DDT and other long-lasting pesticides.

LOCATION: Bays and beaches along the Atlantic and Pacific Coast of the United States and Mexico

DIET: Anchovies, smelt, and other fish

MATURE BROWN PELICAN

ABOUT

The Brown Pelican knows how to make a splash. Grand and sociable, these birds are a familiar sight to many beachgoers. They are excellent swimmers and often fly in a group, called a "pod." As they pass, you may notice them traveling single file over the shore. They often trail fishing boats to snatch seafood caught by humans.

When they do their own fishing, they glide over the water before plunging, head-first, into the waves. That abrupt maneuver may be used to stun their prey. Their expandable bill pouch allows them to scoop up many little fish all at once. In the process, they can collect more than 2.5 gallons (11 l) of water, which they then drain by tilting open their beak.

The Brown Pelican's special scooping technique gives them an advantage in collecting a medley of fishy treats. But other birds are wise to the Pelican's tricks. Gulls, for instance, will hassle Brown Pelicans and steal prey from the bigger bird's pouch.

Fun fact:
Brown Pelican parents don't sit on their eggs. Instead, they use their feet to keep their babies warm, standing right on top of them!

COMMON TERN

Sterna hirundo

APPEARANCE: A sleek, light gray body is topped with a striking black cap. In summer, many of these birds have bright red-orange bills and feet. When it's not breeding season, these areas are dark black.

APPROXIMATE LENGTH: 13″ (33 cm)

VOICE: "keeyur" call

CONSERVATION STATUS: In trouble—climate change and loss of habitat harm these birds

LOCATION: Breeds along beaches, islands, and saltmarshes throughout much of Canada and the northeastern United States

DIET: Squid, leeches, marine worms, and small fish

MATURE COMMON TERN

ABOUT

This dapper seabird lives in crowded, competitive, and often quite noisy colonies. But group life offers many benefits for Common Terns. For example, if a predator approaches, a whole group of Terns will attack. And if one bird in the group discovers food, the rest can join to share the bounty. The Common Tern has a little organ in their beak that helps them separate salt from water so they can stay hydrated even on the ocean.

ROYAL TERN

Thalasseus maximus

APPEARANCE: In their breeding plumage, adults are gray winged and white necked, with sharp black caps that mask their eyes. Outside of breeding season, that cap's color fades. Young Royal Terns are speckled and spotted.

APPROXIMATE LENGTH: 18″ (46 cm)

VOICE: High-pitched "ka-rreek"

CONSERVATION STATUS: While habitat loss remains a danger, they have made a comeback in some parts of their range.

LOCATION: Along ocean and bay shores of much of the United States and Mexico

DIET: Small fish, such as sardines, as well as shrimp, soft-shelled crabs, and other crustaceans

YOUNG ROYAL TERN

ABOUT

These royals are sleek, sophisticated, and sociable. When it's time for Royal Terns to pair up, males present females with a crab or fish treat, bow, and raise their crest feathers. If the offer is accepted, the pair walk together in a circle. Later, when these birds have chicks, Royal Terns gather all of their youngsters together—a bit like a bird daycare—while adults head to the water for fishing. Now and then, a group of normally noisy Terns takes off together in silence. People call this strange behavior "a dread."

GREAT EGRET

Ardea alba

APPEARANCE: White feathers, long black legs, and yellow bills all set apart this Heron species. During the breeding season, the skin around their eyes and bill become minty green and they grow extra-long feathers along their back.

APPROXIMATE LENGTH: 39" (99 cm)

VOICE: Deep croaks and loud squawks

CONSERVATION STATUS: Today, the bird's numbers are stable, though its range may be shifting north. Historically, the Great Egret was part of one of the most significant conservation battles in the United States.

LOCATION: Found in marshy habitat along the coasts of Mexico and United States, some of these birds may breed in the interior of the United States and then winter further south.

DIET: Fish, crustaceans, frogs, snakes, as well as insects and even rodents

GREAT EGRET

But nature-lovers worked together to change laws and protect these animals. Those efforts ultimately saved the Great Egret. Today, their numbers have rebounded and this species is the symbol of the National Audubon Society.

ABOUT

These elegant birds are often observed wading along waterways. Occasionally, you may see the Great Egret pause and patiently wait for a promising fish or frog. When one of those prey items come by, the bird strikes with a fierce jab of its bill.

This graceful wader has a complex history. In the year 1900, an ounce of the Great Egret's gorgeous long head feathers cost double the price of gold. Women in the United States and Europe sought these delicate plumes for their hats. And because people simply could not get enough of them, hunters and skinners made fortunes from the feather harvest. That trade took a serious toll on the birds, nearly wiping out the species.

SIMILAR SPECIES

In much of the same range as the Great Egret, you can observe a similar, smaller white Heron called the **Snowy Egret.** To tell them apart, look for the Snowy Egret's black bill and yellow feet. Yet another species to watch out for is the all-white variant of the Great Blue Heron (page 54). Uncommon outside of the U.S. state of Florida, the white **Great Blue Heron** has pale yellow legs and beak.

HERRING GULL

Larus argentatus

ADULT HERRING GULL IN NON-BREEDING PLUMAGE

APPEARANCE: This Seagull has yellow eyes and bill, light pink legs, and silvery wings edged in white and black. In the non-breeding season, this bird's head and neck are freckled in brown on white. In breeding season, these areas are bright white.

APPROXIMATE LENGTH: 24" (61 cm)

VOICE: Loud "yucca-yucca-yucca" cry

CONSERVATION STATUS: Numbers in steep decline, even though this species is common and widespread

LOCATION: Many of these birds breed in Arctic areas and then migrate across Canada and the United States to warmer winter grounds along the coasts of the United States and Mexico.

DIET: A little bit of everything—including fish, sea urchins, clams, eggs, insects, and human trash

ABOUT

As familiar as these birds may be, the Herring Gull is full of surprises. They have learned to observe human cues and activities to their advantage. Scientists have found that these gulls can read the direction of our gaze and may not snatch our food if they see us watching. One Herring Gull in Paris was spotted as it floated bits of bread in a pond to lure goldfish that the bird then gobbled up.

When Herring Gulls hatch, they instinctively seek out and peck at a small red spot on the lower part of their parent's beak. Doing that prompts mom or pop to feed the little chick.

ACTIVITY

AVIAN ILLUSTRATION

There are two dozen species of gulls in North America. Their basic shape is similar: a stocky rounded body, long wings, and slightly tipped beak. But these birds show tremendous variation from one to the next.

A great way to improve your birding skills is by drawing different species. As you sketch, pay attention to the overall shape first. Then grab watercolors, markers, crayons, or colored pencils to capture the patterns and colors that set various birds apart. You can use this guide as a starting point for Herring Gulls, then look up other species in extended guides.

This project creates a lovely work of art—and it's a great way to boost your attention to detail. For an extra challenge, try sketching other bird groups, such as Herons and Egrets, or colorful songbirds. You can even look up photos of young and old, male and female, breeding and non-breeding birds, to create a comprehensive guide.

ROSEATE SPOONBILL

Platalea ajaja

APPEARANCE: Long spoon-shaped beak, a light greenish head, crimson eyes, white-feathered neck, and a brilliant pink body with equally bright legs distinguish this bird. Chicks are pink-skinned with white feathers, young birds are pale pink and adults are vivid rosy colored.

APPROXIMATE LENGTH: 30" (76 cm)

VOICE: Croaks and clucks

CONSERVATION STATUS: These birds were nearly wiped out of the United States in the late 1800s when their brilliant feathers were prized fashion accessories. In the last 100 years, their numbers have slowly grown. Unfortunately, their habitat is very vulnerable to human activity.

LOCATION: Found in coastal marshes, lagoons, mudflats, and mangrove forests along the Gulf of Mexico and the Pacific coast of Mexico, these birds are also year-round residents of points further south and the Caribbean.

DIET: Small fish and crustaceans, as well as beetles and slugs

ROSEATE SPOONBILL

ABOUT

The glorious Roseate Spoonbill is a vision in pink. Much like the American Flamingo, they owe their gorgeous color to their diet. Many of the crayfish, shrimp, and crabs that these birds consume carry pigments that in turn give these birds their rose-bright colors.

If you're lucky enough to observe these birds in their wetland habitat, you may see several all together. The Roseate Spoonbill lives in flocks and will amble along in groups, often with their bill underwater. They slide this unusually shaped beak back and forth to feel for prey and sift through mud or sand.

These sociable birds have interesting ways of impressing one another. For instance, when male Roseate Spoonbills are trying to impress females, they present them with a stick that they shake in their bills.

This striking bird is unique in many ways. Worldwide, there are only six species of Spoonbill. The Roseate Spoonbill is the only one found in North or South America.

Fun fact:
The scientific name for this bird combines the Latin word for spatula—*Platalea*—and the word for pink—*ajaja*—used in the Tupi language family, spoken by Indigenous communities in the Amazon rainforest of South America.

SANDERLING

Calidris alba

APPEARANCE: In fall and winter, these birds are soft gray above and white below. In late spring and summer, however, their breeding colors will show on mature birds, transforming their head, neck, and back with a rich pattern of red-brown, white, and black. Their black beak and legs are a year-round constant.

APPROXIMATE LENGTH: 7.5" (19 cm)

VOICE: Chatters and "Kip!" call

CONSERVATION STATUS: In serious decline

LOCATION: These birds breed in the Arctic tundra then travel south along the Atlantic and Pacific coastlines of North and South America.

DIET: Sand crabs, marine worms, and other invertebrates

SANDERLING IN NON-BREEDING SEASON PLUMAGE

ABOUT

From fall through spring, you'll see this shorebird poking its beak into sand and mud along temperate and tropical beaches. It is collecting food stirred up by the passing water. Sanderlings rush up just as a wave recedes, then hurry away before another comes. These birds are part of a larger group, called Sandpipers, who use their tweezer-like bills to gather food from the water's edge.

SNOWY PLOVER

Charadrius nivosus

APPEARANCE: These little birds have a white belly and tan-brown head and back. During breeding season, adults have a black spot on the front of their head, a black patch beside each eye, and another on either side of their neck.

APPROXIMATE LENGTH: 6" (15 cm)

VOICE: Sweet "tur-weet!" call

CONSERVATION STATUS: In serious decline in several areas, particularly as people alter their habitat

LOCATION: Along the Pacific Coast from the United States through Mexico, along the Gulf of Mexico coastline, and in the salt flats and alkaline lakes of the interior United States

DIET: Crustaceans, marine worms, mollusks, and insects

SNOWY PLOVER IN NON-BREEDING SEASON PLUMAGE

ABOUT

Skittering along the sand, the Snowy Plover can look like a delicate puff of feathers on quick, toothpick legs. They forage for food by tapping a foot quickly on the ground, creating vibrations that cause worms to rise to the surface. They also rush through washed-up kelp, snapping at flies as they go.

Sometimes, after eggs have hatched, females leave the male to raise their chicks without her. Meanwhile, the female seeks out a new partner to start another family.

WHITE IBIS

Eudocimus albus

APPEARANCE: A curved beak, a red face and legs, and a white body distinguish this species. When these birds are young, they have patches of grayish-brown that lighten up with age.

APPROXIMATE LENGTH: 25″ (63 cm)

VOICE: Grunts and honks

CONSERVATION STATUS: Their numbers in certain areas are dropping although their range may be increasing

LOCATION: Found in swamps, marshes, and brackish or saltwater wetlands along the Gulf Coast of Mexico as well as the Pacific Coast of Mexico and throughout the Caribbean

DIET: Fish, frogs, snails, newts, snakes, and lots of crustaceans

ABOUT

The striking White Ibis is easy to identify because of its curved beak. But these wading birds don't start life with that distinctive feature. The White Ibis hatches with a straight bill that only starts to curve around 14 days of age. They use this beak to reach around in muddy or sandy waters and feel for prey. Once they find something promising, they pinch, pull up, and gulp down their meal. Their beaks also have the benefit of being strong enough to crack open crunchy crustaceans.

ADULT WHITE IBIS

Like many of the birds in this book, there's evidence that White Ibises have learned to adapt to human activity. In the southeastern United States, for example, these birds are taking up residence in city parks more often and developing a taste for breadcrumbs and popcorn. But feeding the birds in this way is a bad idea. Research suggests that birds who do this are at greater risk for catching and spreading disease.

Fun fact:
Male White Ibises show off to females in several interesting ways. They will fly up and down in spiral formations as a group, clean their feathers, and nibble twigs with their bill.

WOOD STORK

Mycteria americana

WOOD STORK

APPEARANCE: This long-legged bird's body is mostly white, with black feathers edging their wings. Young Wood Storks have feathers on their necks that gradually drop away, eventually revealing a dark, bald head and neck as adults. In addition, young birds have light beaks that darken with age.

APPROXIMATE LENGTH: 40" (1 m)

VOICE: Mostly silent but will occasionally croak

CONSERVATION STATUS: In the United States, these birds are not abundant and in the late twentieth century, habitat loss seriously hurt this species. As a result, the Wood Stork was on the endangered species list in the 1980s. The good news is that this bird's numbers are gradually growing and its range has begun to expand.

LOCATION: Swamps and saltwater shorelines of the southern United States and Mexico

DIET: Mullet, minnows, and other fish

ABOUT

Wood Storks nest together in large colonies along with Egrets and Herons. One tree may be a home to two dozen Stork nests, each built of sticks and twigs. To keep their young cool, a Wood Stork will occasionally spit up water over their chicks. If a predator or other intruder—like a human—comes too close to their nests, they will fluff up their feathers and clatter their bills in warning. The Wood Stork is the only Stork species native to North America.

KEEPING WATERWAYS CLEAN

Many birds in this chapter face conservation challenges, including loss of food to overfishing and overexposure to dangerous pollution. By some estimates, 99 percent of ocean-bound birds will have swallowed plastic by the year 2050. Meanwhile, climate change is transforming many waterways.

The good news is that there are many ways you can make a difference.

Three things you can do:

1. **Spread the word!** Talk to your friends and family to make them aware of what's happening. That conversation can be a powerful starting point for change. Your family can buy sustainably farmed seafood, for instance, or symbolically adopt a bird through Project Puffin to support seabird protection.

2. **Reduce, reuse, and recycle.** Plastic trash is a problem for our whole planet. But a lot of simple strategies help. For instance, choose reusable water bottles, straws, and food containers. Opt for paper banners and decorations instead of latex balloons.

3. **Help clean.** Litter can create a choking hazard for wildlife or break down to release toxic chemicals. If you live near a park or beach, volunteer to help clean up local natural areas.

Arid-Climate Birds: Desert Dwellers

CACTUS WREN

Campylorhynchus brunneicapillus

CACTUS WREN

APPEARANCE: Like many Wrens, this bird has a thin, curved bill, and long tail. They come speckled in brown on white with a white stripe above their eyes.

APPROXIMATE LENGTH: 7″ (18 cm)

VOICE: Scratchy, raspy call—a bit like a car that just won't start

CONSERVATION STATUS: These birds are in serious decline, particularly in regions where human populations have grown and transformed the desert landscape.

LOCATION: Look for them amidst thorny plants in low, dry habitats of the southwestern United States and much of Mexico.

DIET: All kinds of insects as well as lizards, spiders, berries, cactus fruit, seeds, and nectar

ABOUT

Is this the ultimate desert denizen? Maybe. The hardy Cactus Wren can be spotted (and heard) singing atop yucca and cacti plants. Unlike most birds, which only use a nest during breeding season, these Wrens sleep year-round in large weed-and-grass nests. That strategy may help when temperatures drop overnight. The Cactus Wren also has a trick to survive with little water: they get most of their liquids from fruit and bug juices.

GAMBEL'S QUAIL

Callipepla gambelii

MALE GAMBEL'S QUAIL

APPEARANCE: These birds have a characteristic head plume made up of several dark, upright feathers. Males and females are largely gray with tan bellies and brown wings accented in white. Males also have a black belly patch, a black-and-white face mask, and a reddish-chestnut crown.

APPROXIMATE LENGTH: 9″ (23 cm)

VOICE: Chip calls and "puk-kwaw-cah" song

CONSERVATION STATUS: Their numbers may be declining in certain areas.

LOCATION: Near water sources within deserts of the southwestern United States and parts of western Mexico

DIET: Seeds, leaves, and fruit such as mistletoe berry and cactus fruit

ABOUT

This round-bodied bird with a distinctive "topknot" scrapes and pecks its way through arid landscapes. It rarely flies and spends much of its time low to the ground, seeking out food. The Gambel's Quail sometimes meets and mates with the **California Quail** in regions where coast and desert meet in the southwestern United States.

GREATER ROADRUNNER

Geococcyx californianus

APPEARANCE: Much of their body is brown streaked on tan. They have a distinct shape, including a crest, long legs, and long tail feathers. During the breeding season, adults have a bright blue, white, and red face patch next to their eyes.

APPROXIMATE LENGTH: 21" (53 cm)

VOICE: "Co-coo-coo" song like a dove, bill clacking noises, and a yipping call that may have inspired the Warner Brothers' "Meep! Meep!"

CONSERVATION STATUS: Decline in some areas though not a species of concern overall

LOCATION: Throughout dry and scrubby territory in much of the southwestern United States and Mexico

DIET: Millipedes, lizards, scorpions, snakes, insects, other birds—these tough birds will even eat leftovers killed by another animal

ABOUT

In the United States, the Greater Roadrunner is a symbol of the southwest. In addition, these birds are important in the beliefs of many Native American communities. The Pueblo tribes, for example, use the Roadrunner's footprint to ward away evil.

GREATER ROADRUNNER

Fun fact:
Roadrunners sometimes swallow snakes slowly. You may see them wandering around, gulping a little bit down at a time.

These desert birds start their day by sunbathing. Nighttime is often very cool in dry environments, so warming up first thing is a must. The Greater Roadrunner lifts some of its feathers, exposing black skin that quickly absorbs heat. Later in the day, they have another trick for cooling off. The Greater Roadrunner has a bald, feather-free patch on their chin that they can flutter to release heat and bring down their temperature.

Like some other desert species, these birds collect water from the liquids in their prey. The Greater Roadrunner is a particularly daring diner. They will munch on poisonous scorpions and lizards. They will even pair up to bring down a rattlesnake, with one bird distracting the reptile and the other going in for a surprise attack.

Although they are not great flyers, the Greater Roadrunner is without question a track and field star. These feathered racers can reach speeds of 20 miles (32 km) per hour, faster than a human. (Though not as fast as their cartoon nemesis, the coyote.) These quick-reflexed birds can also leap into the air to snag a bat or hummingbird.

DINOSAURS AND BIRDS

In the deserts of Wyoming, scientists have uncovered bird fossils that may be 50 million years old. These fossils are the delicate remains of *Calciavis grandei*. They were feathered, hollow boned creatures about the size of a chicken.

Although this landscape is a dry environment today, it has transformed significantly over time. When these prehistoric birds were alive, this region was tropical and lush, with waterways full of fish, plants, and turtles.

The story of birds goes back well before *Calciavis grandei*. In fact, birds are living dinosaurs.

FOSSIL HUNTERS

As the survivors of a massive extinction event, modern birds help scientists who study fossils learn more about the prehistoric world. As one example, researchers can compare the anatomy of living birds with long-lost creatures.

Looking at skeletons, for instance, scientists have determined that the famous winged dinosaur called *Archaeopteryx*, which lived 150 million years ago, probably couldn't fly very well. Instead, this feathery dinosaur may have spent more of its time dashing along the ground—a lot like a Roadrunner.

A STORY OF SURVIVAL

About 66 million years ago, a massive asteroid hit Earth. Not only did it leave a huge dent on our planet—a 93-mile (150 km) crater off the coast of Mexico—it kicked off a wave of extinctions that knocked out most of the dinosaurs living on land at that time. But some species survived, including several from a group of dinosaurs called theropods.

Once there were many varieties of theropod. These dinosaurs included two-legged meat eaters such as Tyrannosaurus rex and velociraptors. But only a few species survived the asteroid. Those that did were small, feathered, and beaked. They were avian dinosaurs, also known as birds.

ARCHAEOPTERYX

PYRRHULOXIA

Cardinalis sinuatus

MALE PYRRHULOXIA

APPEARANCE: These birds are largely gray-brown with a curved, yellow beak. Males have red highlights on their crest, wings, face, tails, and stomachs. Females have red highlights on just their crest, wings, and tail.

APPROXIMATE LENGTH: 8" (20 cm)

VOICE: Sharp chirped song much like a Northern Cardinal

CONSERVATION STATUS: Loss of habitat has hurt these birds in many areas though their overall numbers seem stable

LOCATION: Dry canyons and groves of mesquite trees in Mexico and certain U.S. states on the Mexican border

DIET: Berries, cactus fruit, seeds, and insects

ABOUT

Looking and sounding a bit like the Northern Cardinal, Pyrrhuloxia live in mesquite savannas and scrublands. (Their name, by the way, is pronounced "peer-uh-LOX-ee-a" and can be traced back to the Greek words "pyrruos" meaning "flame-colored" and "loxuos" meaning "crooked.") They collect fruits, insects, and seeds from the ground, using their strong beak to crunch up insects and seeds. They've even been known to chow down on saguaro cactus flowers.

Surviving in deserts requires species to be tough, which may help to explain why paired Pyrrhuloxia are very territorial. Both birds will defend their patch throughout the breeding season.

VERMILION FLYCATCHER

Pyrocephalus rubinus

APPEARANCE: Mature males are crimson red with a fan of dark brown tail feathers, dark wings, and a dark mask over their eyes. Females, meanwhile, are understated in soft browns and grays, with a red-orange lower belly and white chest.

APPROXIMATE LENGTH: 5" (13 cm)

VOICE: "peet-a-weet" call

CONSERVATION STATUS: Some decline is occurring in the U.S. population of this species

LOCATION: These birds are abundant in arid zones and scrublands across the southernmost United States and throughout Mexico.

DIET: Winged insects, including flies and wasps

VERMILION FLYCATCHER

ABOUT

This fiery little Flycatcher perches on shrubs and fences to look for passing insects. Once a target is acquired, this nimble bird will chase its prey and pluck it from the air. Perhaps as a way to show-off their skills, during the breeding season, male Vermilion Flycatchers capture and present butterflies and shining beetles to females.

6

Birds in Fields, Plains, and Prairies: Wide-Open Spaces

AMERICAN KESTREL

Falco sparverius

APPEARANCE: Females and young kestrels have brown upper wings, striped with black. Adult males have more blue-gray coloration on their wings. Males and females have striking black sideburns along their face.

APPROXIMATE LENGTH: 10" (25 cm)

VOICE: Sharp, shrill "klee-klee-klee"

CONSERVATION STATUS: Though widespread, this bird is in decline in some areas.

LOCATION: Across North America—except the Arctic—look for them on a perch with a view, such as a telephone pole by a field or road.

DIET: Mainly large insects, including grasshoppers and moths, but also birds and rodents

FEMALE AMERICAN KESTREL

MALE AMERICAN KESTREL

ABOUT

The American Kestrel may be the smallest Falcon on the continent, but these predators are fierce. Some even hunt in stadiums, making the occasional baseball game cameo. Their colorful feathers serve a purpose. The patches of black on the back of their head look like eyes, fooling Red-tailed Hawks and other big birds that hunt them into thinking they've been spotted.

Fun fact:
American Kestrels sometimes specialize in hunting habits, pursuing just one kind of insect, for example. Talk about picky eaters!

WHAT IS A "RAPTOR"?

Birds of prey, or raptors, are highly specialized hunters. This varied group includes owls, Eagles, Hawks, and Falcons. They share several distinctive features, including strong, curved feet called talons and sharp hooked beaks. Vultures are sometimes included in this group, although their approach to finding meaty meals is very different from other raptors (see 81).

In many raptor species, females are larger than males. Scientists have several ideas about why this occurs. For one, that difference could allow mother raptors to better defend their nest from intruders. Or it may allow males and females to develop different hunting strategies to bring varied food home to their young.

BARN OWL

Tyto alba

APPEARANCE: These owls have a white heart-shaped face. They have light coloration, accented in red-brown and black markings. Females are more colorful than males, with reddish spots on their chest.

APPROXIMATE LENGTH: 14" (36 cm)

VOICE: Hisses, shrieks, and clicks—no hoots from this owl

CONSERVATION STATUS: Declining in some areas, stable in others; people can put up nest boxes to support these birds

LOCATION: Grasslands from the Atlantic to Pacific coasts of the United States, and throughout Mexico

DIET: Small mammals including voles and shrews, as well as the occasional insect, frog, lizard, fish, or bird

FEMALE BARN OWL

Female Barn Owls are more colorful than males. Scientists think that the female's red chest spots serve a purpose. Birds with more of these spots are generally healthier than those with less. In other words, female colors may signal to males that they would make good mates.

ABOUT

The Barn Owl moves like a phantom. These birds are generally solitary and nocturnal, meaning they are most active at night—though you may see them hunting at dusk and dawn too. Their thick layers of feathers soften any sound they make so that they move in silence.

Being quiet gives them the element of surprise. It also helps them hear their prey. In the dark of night, the Barn Owl relies on its ears to detect rodents. In fact, a Barn Owl can hear movement that is 30-feet (9 m) away. That's like hearing the footstep of a mouse at the other end of a school bus.

The owl's distinctive head and face shape sharpen their hearing and help them figure out where a noise is coming from. To further triangulate sounds, they can turn their head around three-quarters of the way.

Fun fact:
These birds are widely spread around the world and can be found on six continents.

BARN SWALLOW

Hirundo rustica

APPEARANCE: A forked tail, iridescent blue back, brick-red throat, and tan underside all distinguish this Swallow species.

APPROXIMATE LENGTH: 7" (18 cm)

VOICE: Squeaky "vit-vit" chatter

CONSERVATION STATUS: Significant decline in Canada and other regions

LOCATION: Look for them swooping to catch bugs over marshes, meadows, and fields across most of the United States, Canada, and Mexico throughout the summer.

DIET: Mostly flying insects, including house flies, horse flies, beetles, wasps, moths, and more

ABOUT

The Barn Swallow is colorful and charismatic. Its distinctive tail and arced wings give it a beautiful silhouette. When these swallows are young, their tail is initially fairly short, then grows to a more pronounced fork in adulthood. These birds breed in North America and winter over in Central and South America. Although the Barn Swallow used to build its cup-shaped mud nest inside of caves, they now use all sorts of human structures, including garages, bridges, porches, and—as their name suggests—barns.

Fun fact:
The Barn Swallow is the most widely distributed Swallow in the world.

BARN SWALLOW

CELEBRATE SWALLOWS, SWIFTS, AND MARTINS

Several species of bird that hunt insects in flight share special characteristics. Swallows, Swifts, and Martins, for example, have long, narrow wings and aerodynamic bodies that allow them to swoop and swerve sharply as they chase their prey.

Because they are such nimble hunters, some farmers encourage these birds to live near their fields, acting as pest control. Homeowners welcome these birds with special nesting boxes, gourds, and towers.

One of the most distinctive swallow conservation projects involves the **Purple Martin,** which is in serious decline in some areas. These birds look a bit like a beefy, darker variety of Tree Swallow, without the white underside (see page 43). They nest in groups called colonies and people put up multiple houses together for these birds. The result is almost like a hanging Martin neighborhood! In some parts of the country, local Audubon chapters organize seasonal Purple Martin parties to gather and watch hundreds of thousands of these birds as they settle in for the evening.

BURROWING OWL

Athene cunicularia

APPEARANCE: This little, leggy owl is mainly marked with brown on sandy-white coloration and has striking light eyebrow patches over bright yellow eyes.

APPROXIMATE LENGTH: 9" (23 cm)

VOICE: "Coo-coooo!" song and chatter

CONSERVATION STATUS: Their numbers are dropping in many areas, in part due to the dangers of cars and loss of habitat. They have also suffered because of efforts to remove prairie dogs and other burrowing mammals, which otherwise create extensive tunnels that offer perfect nesting areas for this owl.

LOCATION: Open grassland, prairies, and even deserts in the western United States and much of Mexico; they also live year-round in parts of Florida

DIET: Beetles, shrews, grasshoppers, earthworms, scorpions, lizards—a whole range of small prey

BURROWING OWL

ABOUT

Burrowing Owls are fascinating characters. They cut quite a striking figure, emerging from their tunnels into the bright sunshine and offering a little head nod. Some people call these owls "howdy birds" because of that greeting-like gesture.

These birds are capable diggers. They can use their beak to excavate into the earth and then their long legs kick dirt piles behind them. Despite those skills, they benefit greatly from other digging critters, like ground squirrels, marmots, and even tortoises. Some people have created artificial burrows to give local Burrowing Owls a boost.

Tunnels offer a safe place for building a nest and storing food. One owl in Canada, for instance, saved up more than 200 rodents for future snacking in a burrow. These owls also have some interesting housekeeping practices, like placing items along the entryway to their tunnel. For example, some Burrowing Owls arrange little balls of animal dung. That poo attracts delicious beetles to the owl's doorstep.

People have also noticed Burrowing Owls arranging human trash, such as scraps of paper, shiny tinfoil, and bottle caps, just outside their burrow. Though people are not sure what the owl is up to, it's possible these birds are simply decorating their abode.

Fun fact:
Young Burrowing Owls will play by jumping on each other, practicing a classic hunting maneuver they will use as adults.

EASTERN KINGBIRD

Tyrannus tyrannus

YOUNG EASTERN KINGBIRD

APPEARANCE: A white belly and throat offer high contrast to this bird's charcoal black head and wings. The regal Eastern Kingbird also has a reddish "crown" of feathers atop its head—though that cap is not easy to see.

APPROXIMATE LENGTH: 8" (20 cm)

VOICE: Loud "kit-kit-kitter-kitter" calls

CONSERVATION STATUS: Numbers decreasing though still widespread

LOCATION: Don't let the "Eastern" name fool you—this bird summers in much of Canada and the United States. It winters in South America's Amazon Rainforest.

DIET: Insects in summer and fruit over winter

ABOUT

The Eastern Kingbird has a reputation for being a tyrant. (That's what inspired the bird's Latin name.) They are indeed highly territorial. These bold birds are not afraid to hassle Hawks, Herons, and even squirrels that get too close. But in winter, they have a much more relaxed lifestyle. They live in flocks and swap out their summer diet of ants and beetles for berries and tropical fruit in South America.

GREATER SAGE-GROUSE

Centrocercus urophasianus

A MALE GREATER SAGE-GROUSE PUFFS UP HIS FEATHERS IN DISPLAY

APPEARANCE: This substantial bird has a mix of gray-brown feathers and a black belly. Females have tan cheeks and white above their eyes. Males have black heads and collars over a white chest.

APPROXIMATE LENGTH: 25" (63 cm)

VOICE: Varied calls including a warning "wut"

CONSERVATION STATUS: These birds rely intensely on their habitat and its plants—the loss of that land and arrival of other flora have put Greater Sage-Grouse numbers in serious decline.

LOCATION: Grasslands in the western United States

DIET: Leaves and buds from the sagebrush plant are a staple, though they also eat insects and other plants

ABOUT

Every spring, lucky nature lovers on the sagebrush plains can witness an amazing spectacle. The male Greater Sage-Grouse begins its display. They puff out their chest—displaying large yellow air sacs—fan their tails like starbursts and begin to bubble over in whistles and little whooping noises. Multiple males will dance and strut together, hoping to catch a female's eye. Often just one or two males in the group will be chosen for mating.

LOGGERHEAD SHRIKE

Lanius ludovicianus

APPEARANCE: A largely gray upper body is accented with black wings, tail feathers, and a mask. These birds have a large head with a sharp, curved beak.

APPROXIMATE LENGTH: 9″ (23 cm)

VOICE: Trills, buzzes, and chattering

CONSERVATION STATUS: Their numbers are in steep decline in many areas

LOCATION: Summering in southcentral Canada and the northwestern and central United States, this bird can be seen year-round in much of the southern half of the United States and in Mexico.

DIET: Lizards, small mammals, and other birds

ABOUT

The Loggerhead Shrike is a powerful predator. Their young practice hunting by chasing each other and pecking at inanimate objects. As adults, they can spear prey with their sharp bill—and will store food for later by impaling it on a branch or bit of barbed wire. When they catch poisonous toads or butterflies, they wait a few days before eating these meals so that the toxic chemicals lose power. The Loggerhead Shrike may be small, but it is strong and can carry prey as big as its own body.

LOGGERHEAD SHRIKE

SCISSOR-TAILED FLYCATCHER

Tyrannus forficatus

APPEARANCE: As adults, these slender birds have elegant long tail feathers and a salmon pink underside. Their upper body is light gray and they have black accents on their wings and tail feathers. Young birds have yellow bellies and shorter tails.

APPROXIMATE LENGTH: 15″ (38 cm)

VOICE: Harsh "kee-kee-kee"

CONSERVATION STATUS: Decline in some areas

LOCATION: Summers in the southcentral United States and winters in Central America

DIET: Insects—including grasshoppers and beetles—and occasional berries

ABOUT

No, that's not a tiny dragon you see swooping past. Admittedly, with its long tail and splash of pink, the Scissor-tailed Flycatcher is otherworldly in its beauty. That showy tail helps these birds twist and turn sharply mid-flight to catch insects. These birds also prize their good looks when seeking out a partner. Males show off their tails with elaborate somersaulting flights to impress a mate.

SCISSOR-TAILED FLYCATCHER

SNOWY OWL

Bubo scandiacus

APPEARANCE: Bright yellow eyes, black bills and feet, and a largely white body make this owl distinctive. Some of these birds have brown or black spots and stripes. Females, which are larger than males, are often more speckled. And young Snowy Owls are often darker than older ones.

APPROXIMATE LENGTH: 25″ (63 cm)

VOICE: Whistles, hisses, and deep hoots

CONSERVATION STATUS: Climate change is transforming the Arctic areas that this species relies on, which may be contributing to the owl's decline across the continent.

LOCATION: These birds breed in the Arctic tundra and winter in parts of Canada and Alaska. Occasionally, they're spotted further south, in the United States—a rare but treasured sight.

DIET: Lemmings and other rodents are a favorite meal—though this owl also eats larger mammals, fish, and birds.

SNOWY OWL

ABOUT

These winter wonders in white plumage are beloved by many. They appear in the Harry Potter books and on cave wall paintings that date back tens of thousands of years. Put simply, these birds have a huge fan base.

Snowy Owls summer in Arctic tundra regions. There, they breed in the open countryside. These owls are not nocturnal. Instead, they take advantage of the long days to hunt lemmings under the midnight sun. Provided those little rodents are plentiful, the owls can find lots of food for their young owlets.

As the seasons change, Snowy Owls journey south. These birds do not follow the same migration routes from year to year, instead they travel to different places from one winter to the next. Their unpredictable arrivals are called "irruptions." Wherever they land, they tend to draw crowds as people swarm to observe these majestic visitors. In North America, some of these birds have appeared in such far-flung and unexpected locales as Hawaii, Bermuda, and southern California.

Fun fact:
In addition to North America, Snowy Owls are found in parts of Europe and Asia.

SWAINSON'S HAWK

Buteo swainsoni

APPEARANCE: A slender raptor with long wings, the Swainson's Hawk comes in a few different color patterns. Some have a reddish chest over a light belly and dark brown feathers above. Others are dark all over. All have yellow feet and yellow beaks tipped in black.

APPROXIMATE LENGTH: 20″ (51 cm)

VOICE: High-pitched "Kreee" call

CONSERVATION STATUS: Once in serious decline, the population of this species has grown in many areas.

LOCATION: In spring and summer, they can be found across grasslands and open agricultural or pasture areas of the western United States; then they migrate to South America for the winter.

DIET: Insects such as grasshoppers and caterpillars—which it will sometimes chase on the ground—as well as reptiles, rodents, and rabbits

SWAINSON'S HAWK WITH DARK COLORATION

ABOUT

Occasionally, you may spot a whole group of raptors swirling together in a group called a "kettle." A kettle of Swainson's Hawk's is an especially exciting spectacle. That's because the Swainson's Hawk travels with thousands of fellow birds. They also undertake the longest migration of any North American raptor, journeying all the way to Argentina.

ACTIVITY

CATCH A THERMAL!

Some bird bodies are perfect at effortlessly soaring through the air. Hawks and Vultures, for example, have wide wings that help them catch currents called thermals. A thermal forms when air heats up—for instance, after the sun bakes rocks on a hill or mountainside—and rises. These pockets of hot air lift and propel birds so they can glide along without flapping their wings.

Want to see a thermal in action? For this activity, you need a grown-up helper. The next time someone is baking something in the oven, ask if you can hold a piece of paper up and over the oven door. (Wear an oven mitt for safety.) Ask your helper to crack the door open and watch as the paper lifts, buoyed by escaping hot air.

TURKEY VULTURE

TURKEY VULTURE

Cathartes aura

APPEARANCE: A bare, red-skinned head tops a large, brown-feathered body. Their beak and feet are white. When flying, their wide wings make a V shape and their feather tips spread slightly, like fingers.

APPROXIMATE LENGTH: 30" (76 cm)

VOICE: Hisses and grunts

CONSERVATION STATUS: Not of concern

LOCATION: Across most of the United States and Mexico

DIET: Carrion, meaning animals that died from other causes

ABOUT

Vultures are often called nature's clean-up crew. That's because they will eat animals hit by cars or killed by illness that many other species won't touch. These birds have several characteristics that help them handle rotten food. Their bare faces, for instance, prevent bacteria from getting stuck in feathers. Their powerful digestive systems fight back against otherwise contagious diseases, such as salmonella. Thanks to their strong stomach, the Turkey Vulture won't get sick or spread germs further.

WESTERN MEADOWLARK

WESTERN MEADOWLARK

Sturnella neglecta

APPEARANCE: These birds are vivid yellow below and streaked with brown, tan, and black above with some white on the sides. Their chest features a striking black patch.

APPROXIMATE LENGTH: 8" (20 cm)

VOICE: Bright, flutelike song

CONSERVATION STATUS: Though common, their numbers may be declining.

LOCATION: Fields and grasslands in the western United States, parts of western Canada and Mexico

DIET: Beetles, crickets, ants, and other insects in summer; grains and seeds in winter

ABOUT

Although male and female Western Meadowlarks look alike to us, they appear different if you—like a bird—can see ultraviolet light! And if you notice these saffron songbirds nearby, tread lightly. They nest on the ground and conceal their eggs with a little grass dome that's easy to miss. Another species, the **Eastern Meadowlark,** looks similar but rarely mixes with the Western Meadowlark.

Woodland Birds: Forest Friends

BARRED OWL

Strix varia

APPEARANCE: These owls have a yellow bill and round, dark eyes, ringed in lighter feathers. The rest of their body is mottled brown and white all over. Their necks and back have a horizontal stripe pattern, leading to the name "barred." Their fronts have more vertical stripes.

APPROXIMATE LENGTH: 18" (46 cm)

VOICE: Hoots "Who cooks for you? Who cooks for you-all?"

CONSERVATION STATUS: Not a species of concern though loss of habitat may be shifting their range.

LOCATION: Across older eastern forests and swamps of the United States; spreading to points north and west

DIET: Mainly mice and other small mammals, sometimes reptiles, amphibians, fish, insects, and birds

BARRED OWL

ABOUT

Hearing or seeing a Barred Owl isn't easy, but it can make a walk through the woods at dusk or dawn extremely rewarding. These birds hunt both day and night but the transition times in the early morning or late evening can be the best time to catch them in action. Look to large trees in mature forests (meaning woodlands where plants have been growing undisturbed for a very long time).

If you're especially lucky, you may encounter a pair of Barred Owls, who often mate for life. These birds select a partner through a courtship that involves hooting duets and bobbing heads side-by-side on a branch. When the pair have chicks, they are very protective. If an intruder gets too close, adults may swoop down, talons first, and attack.

These birds have a complex relationship with other owl species. They do their best to avoid the Great Horned Owl (see page 86), for example, because that bigger species hunts the Barred Owl. Meanwhile, Barred Owls may drive away smaller species, like Spotted Owls, as their own range expands.

Fun fact:
Fossils show that Barred Owls have lived in parts of what is now Canada and the United States for more than 10,000 years.

COMMON RAVEN

Corvus corax

APPEARANCE: Big black birds with prominent, slightly curved beaks

APPROXIMATE LENGTH: 25" (63 cm)

VOICE: Deep croak

CONSERVATION STATUS: In the past century, these birds have begun to reclaim a wide territory across the continent.

LOCATION: This adaptable bird can make itself at home just about anywhere, from high Arctic to desert. They are less common in the eastern United States because people pushed them out of these areas.

DIET: A little bit of everything, including beetles and other insects, reptiles, amphibians, eggs, birds, rodents, and human garbage

COMMON RAVEN

ABOUT

The Common Raven is a living legend. These bold black birds have fascinated humans across continents and cultures for centuries. In North America, they appear in the cultural traditions of many tribes from coast to coast. In the traditions of the Haida people living along the Pacific Northwest, the Raven is a curious trickster, a liberator of their people and a fire bringer. More than 2,000 years ago, the Greek storyteller Aesop told stories about people who believed Ravens could tell the future. And in the 1840s, American author Edgar Allan Poe wrote his most famous poem, "The Raven," based in part on stories of author Charles Dickens' talkative pet Raven named Grip. Spend a little time with these birds and you'll see why they have inspired such awe.

Ravens are clever and can adapt to human activity. You're as likely to see one dumpster diving by a town as you are in the middle of a forest.

In scientific studies, these birds will nimbly solve complex, multistep problems for a food reward. In the wild, they often work in pairs, using teamwork to get a meal. They can use their beak to gesture to one another, much like humans use hands and fingers.

Their social world is rich and complex. Common Ravens will comfort one of their own if that bird has been picked on by another Raven. They're also attuned to fairness. In experiments where human scientists hand out cheese rewards, these birds notice and remember when people eat snacks themselves instead of sharing.

Fun fact:
Birders have spotted Common Ravens sliding down snow-covered roofs for fun. Young Ravens play with sticks and rocks, dropping and catching them midflight.

DOWNY WOODPECKER

Dryobates pubescens

APPEARANCE: This little woodpecker is white below and dark above, with white patches on their head and spots on their wings. Male Downy Woodpeckers have a small tomato red patch on the back of their head. Their sharp pointed beak is black and framed in white feathers. Along the Pacific Coast, these birds have a brownish-gray underside.

APPROXIMATE LENGTH: 6″ (15 cm)

VOICE: Whinnies and high "pik" calls

CONSERVATION STATUS: Not a species of concern

LOCATION: Forests and feeders from Alaska and Canada through Florida

DIET: Insects, especially ants and beetles; they also enjoy suet bird feeders

MALE DOWNY WOODPECKER

ABOUT

The Downy Woodpecker is the smallest of all wood-peckers in North America. They are delicate enough that they can even peck open goldenrod galls—little blister-like spheres on the goldenrod plant—to eat the insect eggs inside.

Like many woodpeckers, the Downy's body is perfectly designed for rapping away at trees. They have strong neck muscles, for example, that help them repeat their back-and-forth motion. They also have a spray of little feathers by their nostrils, to block the sawdust that flies as they chisel into bark. By digging deeply into wood, Downy Woodpeckers can excavate a hole for a nest or, with a more shal-low technique, pull away bark to uncover delicious insect delights.

The Downy Woodpecker has a notable look-alike: the **Hairy Woodpecker.** These two species live in many of the same forests. But the Hairy is bigger and more aggressive than the Downy. Some researchers think that the little Downy Woodpecker has evolved to resemble the bigger, tougher species so that other birds keep their distance!

Fun fact:
Downy Woodpeckers drum on wood noisily to communicate with each other. When they dig for food or excavate for a nest, they are much quieter!

GREAT HORNED OWL

Bubo virginianus

APPEARANCE: This large owl has pointy ear tufts. Their bright yellow eyes are framed with brown disks of feathers, edged in black. Their body is largely a mottled gray-brown.

APPROXIMATE LENGTH: 21" (53 cm)

VOICE: Classic "Hoo-hoo-hoo" hoot

CONSERVATION STATUS: Not a species of concern

LOCATION: Forests throughout North America

DIET: Mice, rats, rabbits, and other mammals, as well as some birds, reptiles, amphibians, insects, and even arachnids

GREAT HORNED OWL

YOUNG GREAT HORNED OWL

ABOUT

Few feathered creatures strike as much fear in their neighbors as the Great Horned Owl. These birds can take down other raptors, including Osprey, Falcons, and fellow owls. They've also been known to attack skunks, rabbits, raccoons, cats, dogs, and even porcupines—although they pay a painful price when they snag a quilled creature.

The Great Horned Owl is nocturnal, but you may see them in the early morning or evening as well. Their enormous eyes are well adapted in ways that let them take in extra light in the dark of night. They also have excellent hearing. Their powerful feet can grip and squeeze prey, allowing them to carry off animals much heavier than themselves.

In the autumn, as part of their courtship, Great Horned Owls hoot a duet. This species breeds in winter and very early spring. That schedule may give young owlets extra training time with their parents that they can use to practice hunting.

The Great Horned Owl's striking ear tufts are not totally understood. They are definitely not ears. An owl's actual ears are hidden beneath their feathers. Some scientists think these pointy feathers enhance sound. It's also possible these tufts help owls camouflage in trees or communicate with each other.

ACTIVITY

DISSECT AN OWL PELLET

Want to learn more about an owl's diet? Owls swallow prey whole or in large pieces. That means their digestive system gradually collects bones, fur, and other tissues that can't be broken down. Owls then cough up those bits and bobs in a little mass called an owl pellet. If you walk in the woods, you may spot one of these pellets—or you can order sterilized owl pellets for an at-home science experiment. With tweezers, it's possible to gently break apart a pellet and reconstruct an owl's past menu. You may even be able to rebuild a whole skeleton. Just remember to wash your hands before and after dissection. And clean up your work area with soap and water when you're finished.

NORTHERN FLICKER

Colaptes auratus

APPEARANCE: This woodpecker has a mostly brownish body, speckled with darker spots and stripes. Males have a stripe on either side of their beak called a mustache. In western North America, these mustaches are red and their heads are gray. But in eastern North America, the mustaches are black and both males and females have a bright red crescent on the back of their head. The eastern variety is called the "Yellow-shafted" Northern Flicker, for the saffron color under its wings and tail. The western variety is "Red-shafted," with brick coloration in those same spots.

APPROXIMATE LENGTH: 11.5" (29 cm)

VOICE: Loud rattle and repeating "wicka-wicka-wicka" call

CONSERVATION STATUS: This bird may be declining in areas where other species take their nest sites.

LOCATION: These birds breed in Alaska and much of Canada and can be spotted year-round throughout many U.S. states and parts of Mexico.

DIET: Insects, fruits, and seeds

FEMALE YELLOW-SHAFTED NORTHERN FLICKER

ABOUT

With their colorful variations, Northern Flickers have kept scientists guessing for years as to whether they are one species or many. With Yellow-shafted in the east, Red-shafted in the west, it's probably no surprise to learn that the Great Plains, in the middle of North America, host Northern Flickers that blend characteristics, including orange under their wings and tails!

You may spot these woodpeckers collecting ants from the ground. Sometimes they rub themselves in crushed ant juices, which carry an acid that may keep mites and other little parasites off of these birds.

During the breeding season, the Northern Flicker will dig into dead trees to hollow out a nesting site. Males will also defend their territory from others with a dance-off. Two males stare each other down, then begin to sway and wave their beaks in elaborate loops. They may flash their wings and call out, continuing the dance until one of the birds backs off.

PILEATED WOODPECKER

Dryocopus pileatus

APPEARANCE: A big, mostly black woodpecker with a prominent red crest and white stripes on its face, shoulder, and neck; males have red stripes to either side of their beak called mustaches

APPROXIMATE LENGTH: 17" (43 cm)

VOICE: High-pitched "cuk-cuk-cuk" and telltale beak knocks

CONSERVATION STATUS: Although this bird was once in steep decline, in the past hundred years or so, forests have been increasing in North America and boosting the number of Pileated Woodpeckers.

LOCATION: Across the southern half of Canada, throughout the eastern United States and along parts of the west coast and northern Rocky Mountains

DIET: Carpenter ants are a major component of meals; these birds also eat termites and other insects, plus fruits and nuts.

MALE PILEATED WOODPECKER FEEDS HIS CHICKS

ABOUT

Every year, the Pileated Woodpecker digs into a large tree using its strong beak. After a few weeks of work, the bird has carved out a perfect safe space for laying their eggs and raising young. At the end of the breeding season, they abandon this spot so they can start the whole process again next year. As a result, there is an opening—or cavity—that ducks, Swallows, owls, and even bats can use in future.

Pileated Woodpeckers also peck at dead or dying trees to look for food. In those cases, they leave behind large rectangular patches in the bark, where they've been chipping away in search of ants, beetles, grubs, and more. These foraging projects draw attention from other birds that also eat insects.

Occasionally, an enthusiastic Pileated Woodpecker will do so much digging that it will cause the tree to split.

All of that pecking away at wood looks like it would take a toll on the woodpecker's body. But woodpeckers can repeatedly make powerful, hammer-like blows with their head without getting hurt. In fact, some scientists study Pileated Woodpeckers to learn more about preventing head injuries in humans!

Fun fact:
The Pileated Woodpecker has a barbed tongue that it uses for collecting insect meals.

RED-EYED VIREO

Vireo olivaceus

APPEARANCE: This small, greenish-brown bird has a lighter underside. Their head is crowned in gray and they have a dark stripe over their eyes, framed in white. Their eyes are brown when young and red as adults.

APPROXIMATE LENGTH: 5″ (13 cm)

VOICE: Repetitive whistled phrases, including "Here I am! Where are you?"

CONSERVATION STATUS: Despite declines in past, numbers are now stable

LOCATION: Breeds in forests across southern Canada and throughout northwestern, central, and eastern United States. Western birds move east during their migration journey to South America.

DIET: Many insects—including wasps, caterpillars, and flies—as well as snails, spiders, and berries

ABOUT

A single Red-eyed Vireo can croon continuously for 10 hours. Their songs are short, musical, and many. One male may cycle through 20,000 tunes in a day. Many of these songs sound like little question-and-answer phrases—as though the bird was chatting to himself for hours on end.

Another clue that there are Vireos nearby is spotting a nest within the trees. Females weave little cup-like structures that hang from tree branches. Often this nest is made from a combination of bark, grasses, spiderwebs, and weeds.

ADULT RED-EYED VIREO

ACTIVITY

BECOME A COMMUNITY SCIENTIST!

Put your birding skills to the test through community science projects. Here are a few ways to get involved:

THE CHRISTMAS BIRD COUNT

For more than 100 years, people have been gathering outside in December and January to tally up all the birds they see. The result is a record for scientists and conservationists who want to know where birds are living and how that's changing over time. To join in, contact your nearest chapter of the National Audubon Society.

THE GREAT BACKYARD BIRD COUNT

This global science project happens during four days in February. All you have to do is watch your local birds during any of those days and share your observations through eBird.org or BirdCount.org.

Looking for more? Good news! There are lots of options. Check out Celebrate Urban Birds, NestWatch, and Global Big Days. Your nearest wildlife refuge or conservation society will have even more suggestions.

WESTERN TANAGER

Piranga ludoviciana

APPEARANCE: Adult females are greenish in color, with gray on their back and wings barred in white and charcoal. Adult males in the breeding season have yellow undersides, mostly black wings and backs, and a bright red face. Outside of the breeding season, their red coloration changes to yellow.

APPROXIMATE LENGTH: 7" (18 cm)

VOICE: Question-like "pit-r-ick" phrases

CONSERVATION STATUS: Not a species of concern—numbers may be increasing

LOCATION: Pine forests along the western half of the continent and throughout much of Central America in the winter

DIET: Lots of insects, including bees, wasps, and cicadas, as well as certain berries and other fruits

MALE WESTERN TANAGER

FEMALE WESTERN TANAGER

ABOUT

Bursting from the branches like a little sunrise, the male Western Tanager is a cheering sight in summer. But in the 1800s, some people saw these birds as a menace because of their enthusiasm for cherries. That conflict led farmers to shoot and poison tanagers. Today, however, people have stopped that practice, in part because of laws designed to protect native species.

If you want to attract these showy birds to your yard, you might put out fresh cut orange slices. But be aware that the Western Tanager may be more interested in insects, at least in the summertime. In fact, the male's bright red face feathers likely get their color from their bug-rich diet.

SIMILAR SPECIES

The Western Tanager is part of a colorful bird family. On the eastern side of the continent, you might encounter a **Scarlet Tanager**. Females of that species look a lot like the greenish-yellow Western Tanager. Males, however, offer a more startling contrast. As their name suggests, male Scarlet Tanagers are mostly red in the breeding season, with black wings and tail feathers. Other times of year, males transform into a subtler yellow-and-black ensemble. If you live in the southern United States, you may see **Summer Tanagers** as they hunt for bees and wasps. Males of that species are the only completely red-feathered birds on the continent!

WILD TURKEY

Meleagris gallopavo

APPEARANCE: These big birds have bodies covered in dark brown, iridescent feathers. Males can fan their tales in display. They have bald heads, framed with dangling skin patches called wattles. Females are less showy, have blue-gray heads, and are smaller bodied than males.

APPROXIMATE LENGTH: 44" (1.1 m)

VOICE: Gobbles and clucks

CONSERVATION STATUS: A hundred years ago, these birds were in trouble. Over-hunting and habitat loss brought them near extinction. But today their numbers are rising and they are spreading into more regions.

LOCATION: Found throughout much of the United States and parts of Mexico, you may see them wandering through open forests or parks and suburbs.

DIET: Acorns, berries, grass, insects, lizards, crabs—these birds eat a varied diet!

ABOUT

Let's talk Wild Turkey. Today, many forests, parks, and suburbs are home to flocks of these birds. If they live nearby, you can observe their group dynamics. Wild Turkeys have strict social roles, with some birds in charge and others following. At night, they take a short flight—they don't fly very much—to sleep in tree branches away from predators.

In Native American communities living in the southwest United States, these birds have long been revered. Their feathers feature prominently in clothing and decoration. In parts of Central America, archaeologists have found evidence for tamed Turkeys that were important to religious events.

MALE WILD TURKEY

When Europeans encountered Wild Turkeys, they imported these birds around the world, with some confusing results. English speakers called the bird "Turkey," perhaps because trade routes passing through Turkey (the country) brought game birds to Europe. The French named the bird "dinde," meaning from India. Malaysians use the phrase "ayam belanda" or "Dutch chicken" because colonists from the Netherlands brought Turkeys to their shores. In Portugal, the bird is called "peru," linking these birds to South America.

Fossils, meanwhile, tell us another story. Prehistoric Turkey bones suggest that 5 million years ago, these birds were strutting and gobbling across what's now the southern United States and Mexico.

Fun fact:
Wild Turkeys can swim! They spread their tails, tuck their wings, and kick their way through the water.

GLOSSARY

Anatomy—(noun) the underlying structure of the body

Archaeologist—(noun) researcher who studies human cultures and remains

Breeding Season—(noun) the period of time when birds find mates, lay eggs, and raise their young

Camouflage—(noun) a way of hiding or concealing one's appearance

Carrion—(noun) the abandoned remains of an animal

Cavity—(noun) a small hole or space for building a nest

Coloration—(noun) pattern of color across the body

Conservation—(noun) the effort to protect and preserve a species or habitat

Courtship—(noun) period when a bird is trying to attract another as a mate

DDT—(noun) dichlorodiphenyltrichloroethane is a chemical used to control the spread of mosquitoes

Deciduous—(adjective) a term used for trees that drop their leaves in the autumn

Evolution—(noun) a scientific theory that explains how species change over many generations

Fossil—(noun) ancient remains or impressions from animals, plants, or other living creatures

Habitat—(noun) the place where an animal lives

Hybrid—(noun) the name for an animal whose parents come from two different species

Mate—(noun) a partner during the breeding season; (verb) to reproduce and have young

Native—(noun) plants or animals that have evolved in a particular ecosystem, as opposed to nonnative species, which have been transplanted to a new ecosystem

Nest Box—(noun) a specialized birdhouse custom built for a species that uses a cavity for building a nest

Nocturnal—(adjective) primarily active at night

Plumage, Plumes—(noun) feathers

Predator—(noun) an animal that hunts and consumes other animals

Prey—(noun) an animal that is hunted and consumed by other animals

Range—(noun) the geographic area where a species lives

Raptor—(noun) a meat-eating bird with a hooked beak and sharp talons

Songbird—(noun) a group of birds that perch and have specialized vocal abilities, including, in many cases, the ability to sing

Species—(noun) a group of animals, plants, or other organisms that share a number of characteristics and can reproduce

Suet—(noun) animal fat used in certain bird feeders

Talon—(noun) a specialized curved claw

Temperate—(adjective) a type of climate that is notable for not being too cold or too hot

Territory—(noun) the area that a specific animal maintains for their own hunting, nesting, and other activities

Tundra—(noun) a type of habitat without trees in the Arctic

RESOURCES

Want to learn more about the birds of North America? There are several great resources—many of which were key references for this book—that you can consult!

First, you can find excellent guides to explore species in greater detail. Several websites offer photographs, audio clips, conservation information, and up-to-date range maps for the many birds of North America. Three online resources to explore are the National Audubon Society's Guide to North American Birds (www.audubon.org/bird-guide), the American Bird Conservancy's Bird Library (https://abcbirds.org/birds), and the Cornell Lab of Ornithology's All About Birds (AllAboutBirds.org).

Apps can help you identify birds on the go, including Merlin Bird ID by the Cornell Lab of Ornithology, which includes a feature to help you recognize birds by their calls and songs.

If you are looking for print species-by-species guides with great visual detail, *The Sibley Field Guide to Birds of Eastern North America (Sibley Birds East)* and *The Sibley Field Guide to Birds of Western North America (Sibley Birds West)* are two beautifully illustrated options. For more fascinating bird facts, David Allen Sibley's *What It's Like to Be a Bird* is an amazing place to start.

COMMON LOONS IN THE SUMMERTIME

Last but not least, if you're hoping to learn more about the incredible journeys that birds make, visit the National Audubon Society's Bird Migration Explorer at Explorer.Audubon.org, to discover which species are flying near you and track their travels.

PHOTO CREDITS

ACKNOWLEDGMENTS

This book owes much to many people! Thank you to my editor, Thom O'Hearn, at Quarto, for your patience and guidance throughout the process.

I am also grateful to several people at the National Audubon Society who provided recommendations and helped review this book for accuracy. That list includes Holly Fairall, Chad Witko, Geoff LeBaron, Mara Silver, Connie Sanchez, and Marlene Pantin. Extra thanks go to Jennifer Bogo, for putting my name forward on this project.

Thank you to Sarah Clark, Stacey Parshall Jensen, and Christine Weeber, who reviewed excerpts from this book and offered excellent sensitivity notes.

I am indebted to the photographers who worked with me on this book: MiaDora McPherson, Judd Patterson, Shawn Carey, Bob Royse, Bruce Campbell, Traci Sepkovic, and Deborah Bifulco. Thank you for documenting the natural world and inspiring others to look closely at the birdlife around us.

On a personal note, I also need to thank several cheerleaders, including my brother Alan and my mom. My husband Tim was my sounding board many times over. Thanks for asking me questions and listening to many, many bird stories that began with "Did you know . . . !?" Now that you've read this book, I'm going to need to learn some new facts to amaze and entertain you.

LAND ACKNOWLEDGMENT

This book explores the stories of birds found across a continent called North America. In many Indigenous communities, including those of the Haudenosaunee and Lenape, this landmass is called Turtle Island.

When you pause to admire the birds of this continent, remember that there are many histories and ways of knowing these animals. This continent has been and remains home to peoples with diverse ways of admiring, engaging with, and understanding nature. That knowledge has been passed down through generations since long before Europeans brought their languages or classification systems for wildlife to the continent.

Often, guidebooks put forward just one way of looking at the world. Each species is a wonder not only in terms of its biology, but also its history and cultural significance across communities. Each community's stories are entwined with those of its neighbors.

ABOUT THE AUTHOR

Daisy Yuhas is a science journalist based in Austin, Texas. Her love of birds has led her to follow Swallows in Alaska and Argentina, count Kestrels in Pennsylvania, boat out to Penguin and Puffin colonies, and seek Dickcissels with physicists on the Illinois prairie. She writes for newspapers and magazines, including *The New York Times for Kids* and *Scientific American*.

INDEX